Environment, Science, and Law

Environment, Science, and Law

An Integrative Introduction

James F. Berry, JD, PhD

Bernan Press

Lanham • Boulder • New York • London

Published by Bernan Press
An imprint of The Rowman & Littlefield Publishing Group, Inc.
4501 Forbes Boulevard, Suite 200, Lanham, Maryland 20706
www.rowman.com

86-90 Paul Street, London EC2A 4NE

British Library Cataloguing in Publication Information available

Library of Congress Cataloging-in-Publication Data

Names: Berry, James F., author.
Title: Environment, science, and law : an integrative introduction / James F. Berry.
Description: Lanham : Bernan Press, 2025. | Includes bibliographical references and index.
Identifiers: LCCN 2024059106 (print) | LCCN 2024059107 (ebook) | ISBN 9798892050739 (cloth) | ISBN 9798892050746 (epub)
Subjects: LCSH: Environmental law. | Climatic changes--Law and legislation. | Science and law.
Classification: LCC K3585 .B467 2025 (print) | LCC K3585 (ebook) | DDC 344.04/6--dc23/eng/20241211
LC record available at https://lccn.loc.gov/2024059106
LC ebook record available at https://lccn.loc.gov/2024059107

♾™ The paper used in this publication meets the minimum requirements of American National Standard for Information Sciences—Permanence of Paper for Printed Library Materials, ANSI/NISO Z39.48-1992.

DEDICATION

This book is dedicated to
the Corps of Cadets
at The Citadel,
The Military College of South Carolina

Honor—Duty—Respect

Contents

Acknowledgments

In my experience, the "acknowledgments" is the hardest part of any book to write. This is especially true for me because this book is the culmination of a lifetime of heavy involvement in both environmental science and environmental law. My greatest regret is that I will certainly fail to recognize some of the people who have been important to me. But I'll give it a shot.

First, I want to acknowledge some of the people who have been important to me as a scientist. Beginning with my graduate professors and colleagues at Florida State University in Tallahassee, I am grateful to the late Dr. Ralph Yerger, Dr. Daniel Simberloff, and Dr. D. Bruce Means. Continuing with the University of Utah in Salt Lake City, I thank my graduate committee, especially the late Dr. John Legler, and my friend and colleague Dr. Richard Shine for their support. I also gratefully acknowledge the support given to a young scientist by the late Dr. C. J. "Jack" McCoy and the staff at the Carnegie Museum of Natural History in Pittsburgh, Pennyslyvania.

I was fortunate to spend my entire scientific career at one institution, Elmhurst University in Illinois. It is impossible to thank everybody who supported me there through the years, but my deepest gratitude is to my friends and colleagues from the Biology Department, including Drs. Stacey Raimondi, Paul Arriola, Tamara Marsh, Kelly Mikenas, Earl Meseth, and the late Dr. Jack Jump.

My current academic home is at The Citadel in Charleston, South Carolina, where I have been treated like family since the day I walked in the door. I am grateful to The Citadel administration for giving me the freedom to experiment with the development of an upper-level general education course that combines environmental science and environmental law. This book will be the textbook for that course. I am grateful to my colleagues in the Biology

Department at The Citadel for their support, including Drs. John Weinstein, Kathy Zanin, John Zardus, and Joel Gramling.

My legal career involves an entirely different group of people. My greatest debt of gratitude is to the late Professor Fred P. Bosselman, who was my mentor, professor, colleague, and friend for many years. Fred was a unique blend of brilliance, pragmatism, and humility, and I miss him always.

I also want to acknowledge some important individuals from my law school days at IIT Chicago-Kent College of Law in Chicago. My sincerest thanks go to Professors Dan Tarlock and Stuart Deutsch.

During my career as an environmental attorney, I have benefited from the expertise of many attorneys, including Nancy Stroud, Edward Ryan, and the other attorneys and staff at the then-law firm Burke, Bosselman & Weaver. I have worked with many other attorneys through the years, but I want to give a special thanks to the late Mark Dennison, with whom I wrote two books and countless articles.

I benefited in many ways through my collaboration with the American Planning Association (APA) in Chicago, especially through Jim Schwab. I have always admired the APA's dedication to environmental responsibility and sensible growth.

I also want to acknowledge the support and assistance from Mary Meghan Ryan, my editor at Bernan Press, and Bob Esposito, my consulting editor for over twenty years.

Finally, I thank my two daughters, Andrea and Jennifer. Most of what I know, I learned from them.

Preface

It's easy for me to see the changes that have occurred in our environment since the mid-twentieth century. I grew up enjoying nature in a relatively undeveloped South Florida, with frequent visits to wild places in Texas and the Bahamas. These places seemed pristine and undisturbed to me at the time, but the truth is that they were already experiencing changes at the hands of people. We had already extensively modified the environment for our own uses, like agriculture, housing, and transportation. The changes have continued, and in most cases accelerated. Like many old naturalists, I pity today's young people who will never experience nature as I knew it then. That said, it could have been a lot worse (and was worse in many places). I think there is reason to be optimistic about the future.

The good news is that the last seventy-five years have seen incredible progress in the STEM sciences. We have improved our understanding of ecosystems, atmospheric patterns, and the effects of pollutants on the environment in ways that allow us to observe and assess the characteristics and the degree of changes to the environment. Along the way, we have developed some promising solutions to some of the problems we have encountered. And we have discovered that there is a need for haste. Climate change is a reality and may be an existential threat to humanity. We will consider these things in this book.

The environmental legal system, by contrast, has been much slower to adapt to environmental change. As we will see, the legal system has developed along a ponderous trajectory that is affected by politics and nonscientific factors that tend to be reactive to specific issues rather than proactive. We will explore the environmental legal system through its origins in common law, current laws and regulations that affect specific aspects of the environment, and some international issues and the agreements among countries that support them.

The primary aim of this book is to explore our modern understanding of the natural environment through the lens of the environmental sciences (environmental biology, chemistry, physics, geology, atmospheric science, and engineering) and the environmental laws that regulate how humans use the environment. More specifically, we will examine the profound influence that environmental laws and regulations have on species, ecosystems, and landscapes, and the effects of the regulation of air, water, and land in maintaining biodiversity and ecosystem health. We will do it in a way that should be understandable and relevant for both non-scientists and scientists, non-lawyers and lawyers alike. Our focus will be on the environmental legal system in the United States, but we will make frequent forays into the international scientific and legal systems.

The book is loosely divided into three sections. Chapters 1 through 5 are primarily introductory. Chapter 1 is an exploration of the origins of environmental policy in the United States, and a discussion of the National Environmental Policy Act (NEPA). In chapter 2, we investigate the American legal system and begin to look at how it relates to the sciences of the environment. Chapter 3 is an introduction to the science of ecosystems, an important focus throughout the book. This focus continues in chapter 4 with an examination of the protection of vulnerable species and ecosystems. Chapter 5 is an investigation of the regulatory system for public lands in the United States with a nod to the regulation of ecosystems.

The second part of the book includes chapters 6 through 9, which are an investigation of air, water, and land. We begin with an examination of the relevant ecosystems and continue with an examination of the regulatory mechanisms that address air pollution, water pollution, and the pollution of land.

The third part of the book has chapters on specialty topics that are usually not covered in books like this. Chapter 10 looks at energy production, renewable and non-renewable energy sources, and the relevant regulatory mechanisms. Chapter 11 investigates the effects of military actions on the environment and the regulatory mechanisms that affect them. Finally, chapter 12 looks at global climate change, including its causes and effects on ecosystems, and legal mechanisms to control climate change.

Each chapter in the book begins with traditional learning outcomes and ends with review questions based on the learning outcomes. At the end of each chapter, there is a section called "Case Analysis." This book is not a "casebook" like those used in many law school courses, but I have found that an examination of the interpretation of laws and legal reasoning by courts is helpful to understanding the relationships between environmental science and environmental laws. Hopefully, you will take the time to review the extracts of court decisions at the end of each chapter.

I hope that you will find our journey worthwhile and that you will consider making your own contribution. After all, the world may depend on it!

J. Berry
Charleston, South Carolina

Chapter 1

National Environmental Policy

Chapter Learning Outcomes

At the end of this chapter, the reader will be able to:

- *Distinguish between the conservationist and preservationist movements in the United States and describe their impact on environmental awareness.*
- *Describe the tragedy of the commons metaphor and its relationship to the need for environmental regulation.*
- *Explain the role of the CEQ under NEPA.*
- *Distinguish between substantive and procedural NEPA.*
- *Discuss the requirements imposed on federal agencies by NEPA §102.*
- *Describe the EIS process.*
- *Explain segmentation under NEPA.*
- *Describe legal standing under NEPA.*

1.1 ENVIRONMENTAL AWARENESS IN THE UNITED STATES

The history of environmental awareness in North America is long and complicated.[1] The indigenous peoples of what became the United States are well known for their veneration of nature, but attitudes were very different among many of the European settlers who began arriving in the seventeenth century. For early European settlers, natural resources were for exploitation, not preservation. For example, between the 1780s and 1860s, about 153 million acres of forest were cleared for farms, and another eleven million acres were cut down for industrial logging, mining, railroad construction, and urban expansion.[2] During this same period, hundreds of thousands of acres of native prairie were converted to farmland, wetlands were drained, and rivers were dammed. The number of European settlers continued to increase through the 1800s and 1900s, resulting in millions of acres of land being converted to human uses including agriculture, mining, industry, and settlements.

Despite European attitudes that favored environmental exploitation, the 1870s and 1880s saw an increase in the influence of philosopher-writers who favored different forms of environmental awareness. These writers formed two philosophical camps. Writers like George Perkins Marsh and Gifford Pinchot supported the view that careful management of the environment would ensure plentiful natural resources for the future. This view, which came to be known as the *conservationist movement*, dominated policy decisions in the United States well into the 1960s.

At the same time in the 1870s and 1880s, a second camp, supported by John Muir and others, advocated for a form of natural morality that supported the preservation of large tracts of pristine wilderness. The approach was largely responsible for our system of national parks and wilderness areas. This became the *preservationist movement*, but it lacked the large public support enjoyed by the conservationists. We will discuss the conservationist and preservationist movements in more detail in chapter 4 in the context of protecting vulnerable species and habitats.

Both the conservationists and preservationists maintained small but stable support during the first half of the 1900s. It was during this time that President Theodore Roosevelt championed a series of environmental conservationist actions that often met with resistance from Western members of Congress. Roosevelt established the United States Forest Service in 1905, signed into law the creation of five national parks, and signed the 1906 Antiquities Act, under which he proclaimed 18 new U.S. National Monuments. He also established the first 51 bird reserves, four game preserves, and 150 national forests, with a total area of approximately 230 million acres (930,000 km^2).[3]

The mid-1900s brought the end of World War II and changes in the attitudes of Americans toward the environment. These changes were fueled by many factors that affected much of American society, including:

- Post–World War II affluence in the United States.
- Scientific evidence of rapid, irreversible environmental damage resulting from anthropogenic pollution.
- Awareness of a history of squandering natural resources.
- Evidence of troubling increases in human population growth rates.
- An increase in demands for major political and legal reforms of American society.

Two popular books helped to fuel interest in the environment. First was the publication of Aldo Leopold's famous book *A Sand County Almanac* in 1959 (figure 1.1). A combination of natural history and philosophy, the book is best known for the following quote, which defines Leopold's land ethic: "A thing is right when it tends to preserve the integrity, stability, and beauty of the biotic community. It is wrong when it tends otherwise."[4]

Then, in 1962, Rachel Carson published her popular inspirational book *Silent Spring* that contemplated a world in which songbirds were extinct

Figure 1.1 *A Sand County Almanac.* It was one of the most influential environmental books of the twentieth century. Aldo Leopold, 1949. *Source*: *A Sand County Almanac.* Oxford University Press. Qualifies as fair use under the copyright law of the United States. Wikimedia Commons. https://en.wikipedia.org/wiki/File:Sand_county_almanac.jpg

due to poisoning by the pesticide DDT.[5] The groundswell of support for Carson's book was massive, and it led to a ban on DDT for most uses in the United States. The book also helped to draw attention to the exposure of many forms of wildlife to environmental contaminants like hazardous chemicals.

It's hard to exaggerate the significance of these events on an America that was eager for societal change. However, overcoming the inertia of federal agencies took longer. One major obstacle to reform was that federal agencies in the 1960s were very program-oriented or mission-oriented, with little regard for the environmental effects of their actions. For example, the U.S. Forest Service, an agency of the Department of Agriculture, was charged with managing the nation's forest resources to optimize timber production, with little regard for the environmental impacts of modifications or destruction of forest ecosystems. The Bureau of Land Management (BLM), an agency of the Department of the Interior, managed the nation's public lands in support of mining and cattle grazing, with little attention to environmental impacts.

By the mid-1960s, Congress was ready to support governmental changes at many levels. Federal agencies began to change their attitudes toward the environment. Of significance was the passage of the National Environmental Policy Act (NEPA) in 1969 and the creation of the Environmental Protection Agency (EPA), a Cabinet-level executive agency. We will discuss NEPA and the EPA further below.

Congress began to address environmental concerns in the 1960s, but there was a lack of consensus on how to proceed. Professor Tarlock suggested three alternative forms for environmental legislation at the time:[6]

- Federal agencies could be required to consider the environmental consequences of their actions in the decision-making process (the approach adopted in NEPA);
- Designated natural resource areas could be withdrawn from development and exploitation, or environmental agencies could have a veto on proposed agency actions (the approach used for floodplains and wetlands); or
- Agencies could be authorized to adopt environmental standards and to prohibit development that violates these standards (other environmental laws, such as the Clean Air Act, Clean Water Act, and Resource Conservation and Recovery Act use this approach).

We will see numerous examples of these approaches throughout this book. As we will discover, the period of time from 1969 to the early 1980s saw the passage of at least thirty U.S. environmental laws affecting nearly every aspect of the environment. Since that time, public awareness of environmental issues has continued to increase to the point that current issues like global climate change (discussed in chapter 12) have become rallying cries for governments and organizations around the world.

1.2 THE TRAGEDY OF THE COMMONS, AND THE NECESSITY FOR REGULATION

It is not uncommon for members of the communities that are affected by federal environmental laws and regulations to argue that environmental protection should either not exist or be voluntary.

We can use the metaphor of the Tragedy of the Commons to explain the difficulty of voluntary controls on the use of resources that are publicly owned. In 1968, British ecologist Garrett Hardin published his famous paper *The Tragedy of the Commons.*[7] He described a medieval system in England in which a town would have a publicly owned Town Commons on which residents were free to graze their sheep at will (figure 1.2). The Town Commons would support a certain population of sheep and continue to renew itself as long as the maximum number of sheep (called the **carrying capacity**[8]) was not exceeded. As long as all members of the town community behave cooperatively in an altruistic manner and do not exceed the carrying capacity of the Town Commons, then the system works.

Figure 1.2 A Town Commons in Modern England. Brampton Ash, Northamptonshire, England. *Source*: Wikimedia Commons. Tim Heaton / We are His people and the sheep of His pasture (Psalm 100:3) / CC BY-SA 2.0. https://creativecommons.org/licenses/by-sa/2.0/deed.en

The problem, as explained by the Tragedy of the Commons metaphor, is that individuals may begin to act in their own self-interest to the detriment of the Town Commons. Imagine that one resident of the town wants to increase the size of their sheep herd. If the person adds more sheep to the Town Commons and exceeds the carrying capacity, then the entire Commons becomes degraded, and the entire town will suffer. In other words, the Tragedy of the Commons metaphor explains the breakdown of altruism when individuals behave selfishly and exploit a publicly owned resource for their own benefit. It has been used for over fifty years to explain behavior in fields such as economics, psychology, and ecology.

In environmental issues, the Tragedy of the Commons is used to explain the selfish overuse of the Global Commons, which is the term used for worldwide natural resources that are publicly owned. The idea is that businesses and industries will behave in their own self-interest and exploit the Global Commons in a way that maximizes their profits but depletes the resources in the Global Commons. It is argued, then, that the purpose of environmental laws and regulations is to control the kind of selfish behavior contemplated by the Tragedy of the Commons.

As useful as the Tragedy of the Commons metaphor has been, it definitely has its limitations and detractors. It has been argued that it is an oversimplification, or that it underestimates the impact on the Global Commons, or even that it is a "dubious, right-wing concept."[9] Other writers with a more hopeful interpretation have seen the Tragedy of the Commons as an outlier, and that humans often show cooperation and respect under challenging circumstances.[10]

We will encounter several references to the Tragedy of the Commons in later chapters.

1.3 THE NATIONAL ENVIRONMENTAL POLICY ACT (NEPA)

In response to the factors we discussed earlier Congress passed the first piece of legislation directly designed to protect the environment in 1969, under the Presidency of Richard Nixon. That groundbreaking legislation was the National Environmental Policy Act (NEPA). The Environmental Protection Agency (EPA), an executive agency, was created in 1970 (figure 1.3).

The significance of NEPA to environmental awareness in the United States cannot be overstated. In fact, Professor Rodgers described NEPA as "the most famous statute of its kind on the planet."[11]

The structure of NEPA is different from most other federal environmental laws. NEPA language was intentionally short and vague and lacks the

Figure 1.3 An Agency's Decision Process under NEPA Section 102. *Source*: U.S. Department of Energy. Office of Environmental Management. NEPA Reviews. https://www.energy.gov/em/nepa-reviews

extensive regulatory language of other statutes. Furthermore, NEPA has no instructions for judicial review, which encouraged the development of a "common law" of NEPA, in which courts (judges) would determine how NEPA would be applied to a particular case.[12] However, this approach produced a series of unintended consequences in the early years of NEPA, such as forum shopping,[13] inconsistent court decisions based on the inclination of judges, and a U.S. Supreme Court that extended agency power while lower courts extended judicial power. These concerns have been addressed to varying degrees through the years, but they remain a source of criticism of NEPA.

NEPA Section 2

NEPA §2 contains Congress' Declaration of Purpose for NEPA. The language of NEPA is a remarkable statement of support for the environment by Congress:

The purposes of this Act are: To declare a national policy which will encourage productive and enjoyable harmony between man and his environment to promote efforts which will prevent or eliminate damage to the environment and biosphere and stimulate the health and welfare of man; to enrich the understanding of the ecological systems and natural resources important to the Nation; and . . .[14]

It is clear from the language that a seismic shift in attitudes toward the environment was underway.

Nepa §2 also creates the **Council on Environmental Quality (CEQ)**. The CEQ is an Executive office that coordinates federal environmental efforts in the United States, working closely with federal agencies and other Executive offices to develop environmental and energy policies and regulations. The CEQ has three members, chosen by the president and approved by the senate. One of the three members serves as the Chair of the CEQ. The idea is that CEQ members have training and experience to interpret data and suggest national environmental policy. However, CEQ members are often political appointees and do not always have the environmental expertise that Congress contemplated.

NEPA Section 101

Section 101(a) of NEPA sets forth a national policy to use:

. . . *all practicable means and measures, including financial and technical assistance, in a manner calculated to foster and promote the general welfare, to create and maintain conditions under which man and nature can exist in productive harmony, and fulfill the social, economic, and other requirements of present and future generations of Americans.*[15]

Continuing with its mandate, NEPA §101(b) states that "it is the continuing responsibility of the Federal Government to use all practicable means, consistent with other essential considerations of national policy" to avoid environmental degradation, preserve historic, cultural, and natural resources, and "promote the widest range of beneficial uses of the environment without undesirable and unintentional consequences."[16]

NEPA §101(c) contains the following intriguing statement: "The Congress recognizes that each person should enjoy a healthful environment and that each person has a responsibility to contribute to the preservation and enhancement of the environment."[17] This statement seems to create an obligation on all Americans to preserve and protect the environment. Does this create an "enforceable right" to a protected environment? In other words, can we sue a person (or an agency) who is not preserving and protecting the environment? It appears that no such right exists.[18]

Collectively, NEPA §101 is often called **substantive NEPA** because it seems to create substantive rights and obligations by telling the agencies what the future should look like. However, a series of U.S. Supreme Court decisions has made it clear that NEPA establishes procedural, rather than substantive, obligations on federal agencies.[19]

NEPA Section 102

NEPA §102 is known as **procedural NEPA** because it imposes specific procedural requirements on all federal agencies. The Supreme Court has declared that NEPA is "essentially procedural," so NEPA §102 is the most significant part of the statute.[20]

Section 102(2) imposes several procedural duties on federal agencies. The most prominent of these obligations is that of Section 102(2)(C), which states that:

> *The Congress authorizes and directs that, to the fullest extent possible: . . . (2) all agencies of the Federal Government shall*
>
> *. . . (c) include in every recommendation or report on proposals for legislation and other major Federal actions significantly affecting the quality of the human environment, a detailed statement by the responsible official on:*
>
> *(I) the environmental impact of the proposed action,*
>
> *(ii) any adverse environmental effects which cannot be avoided should the proposal be implemented,*
>
> *(iii) alternatives to the proposed action,*
>
> *(iv) the relationship between local short-term uses of man's environment and the maintenance and enhancement of long-term productivity, and*
>
> *(v) any irreversible and irretrievable commitments of resources which would be involved in the proposed action should it be implemented.*[21]

The details of the process are clarified in CEQ's NEPA implementing regulations.[22] The detailed statement of the environmental impact of an action that must be prepared by each federal agency is known as the "environmental impact statement" (EIS). The EIS requirement is the cornerstone of NEPA. Clauses (i) through (v) enumerate the contents of the EIS, which we will discuss below. Clauses (iv) and (v) address the problems of long-term commitments of resources and their effects on future generations. Unfortunately, these clauses generally do not receive adequate attention either in EISs or in reviewing courts.

In most cases, the NEPA §102(2)(C) requirements are triggered only for every *major federal action significantly affecting the quality of the human environment*. But what, exactly, does that mean? We can parse the statement to see what courts and commentators have told us:

- *Is it Federal?* This generally requires a federal **nexus** (connection). As a general rule, if the agency has the power to influence the outcome materially, then the action is federal.
- *Is it a major action?* The regulations say "any action that: (1) Receives more than $100,000,000 in federal funding; (2) Requires a federal permit to commence the project; or (3) Crosses a boundary between states or an international boundary." The word "major" means "reinforces but does not have a meaning independent of significant."[23]
- *Does it significantly affect the quality of the human environment?* CEQ regulations state that this inquiry is based on "considerations of both context

and intensity," and it lists ten important factors, including the "highly controversial" nature of the impact.[24] Some agencies define "significant" in their own regulations, which is used as guidance by the EPA. Many court cases have resulted from individuals, groups, or government agencies who are disappointed with an EIS decision.

The NEPA §102 Impact Procedure

Once an agency determines that its proposed action meets the requirements of NEPA §102(2)(C), the environmental review under NEPA can involve three different levels of response:[25]

1. **Categorical Exclusion (CATEX)**. Under certain circumstances, a federal agency action may be categorically excluded from a detailed environmental analysis when the federal action normally does not have a significant effect on the human environment.[26] The reason for the exclusion is generally detailed in NEPA procedures adopted by each federal agency. These are usually projects that are often repeated with similar environmental consequences each time, such as most highway and bridge construction, mining operations, and flood control projects.
2. **Environmental Assessment/Finding of No Significant Impact**. If the federal agency determines that a Categorical Exclusion (CATEX) does not apply to a proposed action, then the agency may prepare an Environmental Assessment (EA). The EA determines whether or not a federal action has the potential to cause significant environmental effects. Each federal agency has adopted its own NEPA procedures for the preparation of EAs. Generally, the EA includes a brief discussion of: (1) The purpose and need for the proposed action, (2) alternatives to the proposed action,[27] (3) the environmental impacts of the proposed action and alternatives, and (4) a list of agencies and persons consulted. Based on the EA, the following actions can occur:
   - If the agency determines that the action will not have significant environmental impacts, the agency will issue a Finding of No Significant Impact (FONSI), a document that presents the reasons why the agency has concluded that there are no significant environmental impacts.
   - If the EA determines that the environmental impacts of a proposed federal action will be significant, an Environmental Impact Statement is prepared.

   About 50,000 EIAs are prepared each year, making them the most common agency response by far.[28]

3. **Environmental Impact Statement (EIS).** Federal agencies prepare an EIS if a proposed major federal action is determined to significantly affect the quality of the human environment. The regulatory requirements for an EIS are much more detailed and rigorous than the requirements for an EA.

The agency decision process is summarized in figure 1.3.

Summary of the EIS Process

The EIS process is lengthy, expensive, and very public, making EISs controversial and relatively rare. Only about 500 EISs are prepared each year.[29]

Many EISs are prepared for traditional federal agency actions involving public projects, like the removal of spent fuel rods at a nuclear waste facility by the Department of Energy or building and demonstrating a mobile microreactor for energy production in the field by the Department of Defense. The EPA maintains a database of EISs.[30] Surprisingly, though, many EISs are prepared for projects that are private. For example, in 2019 an EIS was prepared for a project in Mississippi that expanded natural gas liquefaction and export capabilities at a facility owned by Gulf LNG Energy, LLC (GLE), a private corporation. Because the project involved licenses and funding from the Department of Energy, there was a sufficient nexus to a federal agency, and an EIS was prepared.

The EIS process involves several stages. First, the federal agency most involved in the project (called the *lead agency*) publishes a *Notice of Intent* in the Federal Register, which informs the public of the upcoming environmental analysis and describes how the public can become involved in the EIS preparation. The involvement of the public at the earliest stages of the EIS process reinforces Congress' intention that the entire process be public and transparent. This public involvement has been both supported[31] and condemned,[32] but it remains one of the most important aspects of NEPA.

The *Notice of Intent* initiates the *scoping process* during which the federal agencies and the public collaborate to identify the issues and potential alternatives to be addressed in the EIS. A Draft EIS (DEIS) is then published for more public review and comment for at least forty-five days. After the comment period closes, the agencies consider all substantive comments and, if necessary, conduct further analyses.

After the agencies have made all adjustments, a Final EIS (FEIS) is published, which includes responses to substantive comments. The publication of the Final EIS begins a "wait period," in which agencies are generally required to wait thirty days before making a final decision on a proposed action. At this point, the EPA publishes a Notice of Availability in the Federal Register, announcing the availability of both Draft and Final EISs to the public.

The EIS process ends with the issuance of the *Record of Decision (ROD)*, which explains the agency's decisions, describes the alternatives the agency considered, and discusses the agency's plans for mitigation and monitoring, if necessary.

A supplement to a draft or final EIS is prepared when an agency makes substantial changes to a proposed action that are relevant to its environmental concerns, or there is significant new information relevant to the environmental concerns that have a bearing on the proposed action or its impacts. If an agency decides to supplement its EIS, it prepares, publishes, and files the Supplemental EIS (SEIS) in the same way as a DEIS or a FEIS.

An important question is who, exactly, prepares the EIS? NEPA clearly states that the EIS is to be prepared by the agency. However, most federal agencies do not have the expertise or the personnel to prepare an EIS. Federal agencies typically outsource the writing of an EIS to third-party contractors (including lawyers, scientists, and engineers) with expertise in their preparation and in relation to the proposed project.[33] However, the preparation probably cannot be delegated to the "applicant" (usually a private party who stands to benefit from the action), because it would be too self-serving. It is probably acceptable for the EIS to be prepared by the applicant's consultants, though, because it is not thought to be self-serving.[34] Under amendments to NEPA §102(2)(D),[35] preparation of the EIS can be delegated to a state.

EIS Scope and the Segmentation Problem

There have been concerns about the scope of EISs since the earliest days of NEPA. A particular concern regards *segmentation*, or the practice of avoiding the appearance of "cumulative impact" problems by breaking up large projects into several smaller ones. The intention is that these smaller projects can avoid the preparation of an EIS. In order to avoid segmentation problems as much as possible, CEQ regulations require an EIS to discuss: (1) connected actions (closely related and should be discussed in the same impact statement), (2) cumulative actions, and (3) similar actions.[36]

An example of segmentation occurred when state and federal highway authorities prepared an EIS for the expansion of two segments of an interstate highway in the early 1970s. A federal lawsuit by private citizens charged that the EIS for the highway segments ignored the cumulative impact on a state park that was in the middle of the two segments. The court held that this was illegal segmentation and required the highway authorities to complete an EIS that included the effects on the park.[37]

The segmentation issue continues to draw attention in the environmental legal community. In 2020, the CEQ issued revised regulations under NEPA in response to an Executive Order (which was reversed in 2021), which directed CEQ to "ensure that agencies apply NEPA in a manner that reduces unnecessary burdens and delays as much as possible."[38] In other words, the Executive Order directed CEQ to use its statutory authority to interpret NEPA in a way that simplifies and accelerates NEPA review. The 2020 regulations shrink the scope of an EIS by omitting the cumulative effects analysis despite many court decisions reinforcing that NEPA unambiguously requires the cumulative effects analysis in every EIS.[39] Commentators worry that this change would open the door to segmentation and result in unnecessary environmental damage.[40]

NEPA Litigation

We will discuss the legal system much more thoroughly in chapter 2, but let's look at lawsuits as they occur under NEPA. Because this is an action between parties, it is a civil lawsuit, rather than a criminal suit where the State charges a party with committing a crime. For a civil lawsuit to occur, most federal laws contain a statement that specifically allows lawsuits against the government by interested parties. NEPA does not have such a statement, but interested parties can file lawsuits against federal agencies under provisions of the federal Administrative Procedures Act (APA).[41] Most NEPA lawsuits are filed by parties alleging that a federal agency has not complied with their legal obligations under NEPA. Under the APA, these lawsuits may only be filed in Federal court. However, as an alternative, parties are encouraged to submit their suits to the much faster and less expensive Administrative Appeals Board (AAB), an impartial appellate tribunal established by CEQ regulations that is headed by a hearing officer to hear administrative appeals under the major environmental statutes that EPA administers, including NEPA.

NEPA lawsuits are not common, but they tend to follow one of two patterns. The first is a lawsuit in which a party sues a federal agency for not following the mandates of NEPA. The party filing the lawsuit, or *plaintiff*, may be a private citizen but is often an interest group like an environmental organization. The second type of lawsuit is by members of the regulated community who are suing to challenge the validity of a provision of NEPA as it applies to them. Because the law of NEPA is mostly judge-made law, these cases often have a lengthy pathway through the courts, including the appellate system. However, fewer than twenty NEPA lawsuits have reached the U.S. Supreme Court. The rich legal history of NEPA is of little interest to the public, but it has fascinated environmental lawyers for decades.[42]

Before we proceed, it's important to note that not anyone can file a NEPA lawsuit. Under the provisions of the Administrative Procedures Act, a party to a lawsuit must have *standing* to sue. We will address the issue of standing more extensively in chapter 2, but for now it means that the party must be among those affected by an agency decision under NEPA. So, if an agency's action (or inaction) causes harm to a person's interest, then they probably have standing to sue, as long as the harm is not too abstract or remote. For environmental groups, their members may have standing if they suffer injury to their esthetic, conservation, or economic interests, or if members use a wilderness area for recreation, or noise and air pollution would harm members' health or conservation interests. But environmental groups lack standing if their interests are too abstract or remote, or if members do not show personal injury. Animals do not have standing under NEPA.[43]

Once we establish that our party has standing to sue, the next inquiry is which section of NEPA applies. Recall that NEPA §101 creates what seem to be substantive rights and obligations. Can we sue a federal agency for failing to meet the substantive requirements of NEPA? It turns out that we cannot. The U.S. Supreme Court held that a court's only role is to see that the agency considered the environmental consequences of the proposed action. There is no cause of action for violation of substantive NEPA.

May a court reverse a decision in favor of an agency action for failure to follow NEPA's procedural requirements under NEPA §102? Courts have held that NEPA is "essentially procedural."[44] But even if a party sues under §102, they are only likely to prevail if it can be proven that the agency failed to follow its procedural requirements, like preparing a flawed EIS or failing to prepare an EIS at all. A court will not reverse an agency action because the agency's final decision was not in the best interests of the environment. As the Supreme Court held, "NEPA merely prohibits uninformed—rather than unwise—agency action."[45]

Before we conclude our exploration of NEPA litigation, we must inquire about possible remedies. In other words, what do you get if you win? In NEPA litigation under the Administrative Procedure Act, damages are not possible, either monetary or otherwise. Damages may be possible under some *common law* theories that we will discuss in chapter 2, but not under NEPA.

Remedies under NEPA are limited to *injunctive relief*, also known as an *injunction*, which is a remedy that prevents a party from doing certain acts or requires them to perform in a particular way. In a NEPA case, this usually means that the agency must alter its decision on a proposed action, and either reconsider or change the environmental impacts. However, injunctive relief in a NEPA case is almost always temporary. Permanent injunctions are virtually impossible under NEPA.

Since the only real remedy under NEPA is a temporary injunction, some commentators have questioned whether NEPA is even minimally effective.[46] However, the more common (and hopeful) view is that NEPA has saved lives, preserved community integrity, protected endangered species and public land, and saved billions of dollars. NEPA has often been the first and last line of defense against government mismanagement and industry abuse.[47]

NEPA's International Reach

Before we leave NEPA, we should acknowledge that NEPA reaches beyond the borders of the U.S. NEPA §102(2)(F) states that federal agencies must "recognize the worldwide and long-range character of environmental problems."[48] However, there has been concern from the beginning whether NEPA can be enforced where federal international actions are concerned. In 1979, President Carter issued an Executive Order that requires a less demanding procedure (such as an Environmental Assessment) for most federal actions.[49] But many agency actions (like the military) are exempted from the process, and there is no private "cause of action" by plaintiffs wishing to challenge a federal agency's decision. The true effectiveness of NEPA in the international arena is debatable.

Although NEPA's international reach remains a question, NEPA has been an important international role model. More than one-hundred governments of nations around the world have enacted national environmental policies modeled after NEPA.[50] With its requirement for transparency and environmental sensitivity, it remains one of the most important environmental laws on the planet.

We will explore many other environmental laws in forthcoming chapters. As we do, compare the approaches that Congress used in creating the laws, how the laws are structured, and how they are enforced.

* * *

Review Questions

1. Describe the impacts of early European settlers in North America on native forests.
2. Contrast the views of the conservationist movement and the preservationist movement in nineteenth-century America.
3. What are Professor Tarlock's three approaches that the U.S. Congress took in establishing environmental laws in the 1960s and 1970s?
4. Using the Tragedy of the commons metaphor, explain why voluntary environmental protection is seldom successful.

5. Explain why the structure of NEPA promoted the development of a "NEPA common law."
6. Why is NEPA §101 called "substantive NEPA"?
7. Describe the process by which an environmental impact statement is prepared.
8. Under what circumstances can a person sue under NEPA?

Case Analysis

The Fifth Amendment to the U.S. Constitution says that it is unlawful for the government to "take" the property of a citizen for public use without "just compensation." The idea is that you should receive fair market value for your land if, for example, the state takes your property as part of a highway construction. Federal courts have repeatedly upheld the "takings" provision of the Constitution.

The issue becomes less clear, however, in the case of a "regulatory taking." Suppose that a developer owns ten acres of land on the coast of South Carolina and wishes to build several single-family homes on his beachfront property. However, a state law for beach preservation prevents him from building any homes at all. Should the state be required to reimburse him for the $1.2 million dollars that he paid for the property?

These are essentially the facts in the Supreme Court's 1992 decision in *Lucas v. South Carolina Coastal Council.* Lucas paid over $9 million for beachfront property on the Isle of Palms in South Carolina. However, a state law, the Beachfront Management Act, prevented him from building homes on the property and realizing the value of his investment. The following is an extract from the Supreme Court's decision in Lucas. Read the excerpt, and answer the questions below.

> Where the State seeks to sustain regulation that deprives land of all economically beneficial use, we think it may resist compensation only if the logically antecedent inquiry into the nature of the owner's estate shows that the proscribed use interests were not part of his title to begin with. This accords, we think, with our "takings" jurisprudence, which has traditionally been guided by the understandings of our citizens regarding the content of, and the State's power over, the "bundle of rights" that they acquire when they obtain title to property. It seems to us that the property owner necessarily expects the uses of his property to be restricted.
>
> from time to time, by various measures newly enacted by the State in legitimate exercise of its police powers; "[a]s long recognized, some values are enjoyed under an implied limitation and must yield to the police power." And in the case of personal property, by reason of the State's traditionally high degree

of control over commercial dealings, he ought to be aware of the possibility that new regulation might even render his property economically worthless (at least if the property's only economically productive use is sale or manufacture for sale). In the case of land, however, we think the notion pressed by the Council that title is somehow held subject to the "implied limitation" that the State may subsequently eliminate all economically valuable use is inconsistent with the historical compact recorded in the Takings Clause that has become part of our constitutional culture.

Where "permanent physical occupation" of land is concerned, we have refused to allow the government to decree it anew (without compensation), no matter how weighty the asserted "public interests" involved, though we assuredly would permit the government to assert a permanent easement that was a pre-existing limitation upon the landowner's title. We believe similar treatment must be accorded confiscatory regulations, i.e., regulations that prohibit all economically beneficial use of land: Any limitation so severe cannot be newly legislated or decreed (without compensation), but must inhere in the title itself, in the restrictions that background principles of the State's law of property and nuisance already place upon land ownership. A law or decree with such an effect must, in other words, do no more than duplicate the result that could have been achieved in the courts by adjacent landowners (or other uniquely affected persons) under the State's law of private nuisance, or by the State under its complementary power to abate nuisances that affect the public generally, or otherwise.[51]

Questions

1. Justice Scalia, who wrote the majority opinion in Lucas, was known for his flowery language that is not always easy to follow. As you read this excerpt, did the Supreme Court ultimately agree with Mr. Lucas or with the State of South Carolina?
2. Suppose that Mr. Lucas had the option of charging admission to his undeveloped land, rather than building houses, but that would only allow him to recover about 5 percent of what he paid for the land. Would that still be a regulatory taking?
3. The Court mentions the state's "police power," which allows it to regulate private property to protect public health and welfare. Should this allow the state to prevent construction on the beachfront? What did the Supreme Court say?

NOTES

1. There are many excellent scholarly and popular reviews of environmental awareness in the United States. See, for example: Merchant, C. 2007. *American Environmental History: An Introduction*. Columbia University Press; and Fiege, M. 2013.

The Republic of Nature: An Environmental History of the United States. University of Washington Press.

2. Williams, M. 2003. *Deforesting the Earth: From Prehistory to Global Crisis*. The University of Chicago Press.

3. There are many excellent biographies of Roosevelt and his record in environmental conservation. For example, see: Miller, N. 1992. *Theodore Roosevelt: A Life*. William Morrow & Co. and Brinkley, D. 2009. *The Wilderness Warrior: Theodore Roosevelt and the Crusade for America*. HarperCollins.

4. Leopold, A. 1949. *A Sand County Almanac*. Oxford University Press.

5. Carson, R. 1962. *Silent Spring*. Houghton Mifflin Company.

6. Tarlock, A.D. 1972. *Balancing Environmental Considerations and Energy Demands: A Comment on Calvert Cliff's Coordinating Committee, Inc. v. AEC. Indiana Law Journal, 47,* 645–679. See also: Tarlock, A. D. 2010. Environmental Law: Then and Now. *Washington University Journal of Law & Policy, 32,* 1–31.

7. Hardin, G. 1968. The Tragedy of the Commons: The Population Problem Has No Technical Solution; It Requires a Fundamental Extension in Morality. *Science, 162*(3859), 1243–1248.

8. We will discuss the concept of the carrying capacity of a population more fully in chapter 5.

9. Amer, A. 2023. The "Tragedy of the Commons" Is a Dubious, Right-Wing Concept. *Jacobin Magazine*, October 20. https://jacobin.com/2023/10/tragedy-of-the-commons-garrett-hardin-white-supremacy-enclosure-privatization-history. See also Angus, I. 2008. The Myth of the Tragedy of the Commons. *Climate & Capitalism*, August 25. https://climateandcapitalism.com/2008/08/25/debunking-the-tragedy-of-the-commons/.

10. For example, see: Ostrom, E. 1990. *Governing the Commons: The Evolution of Institutions for Collective Action*. Cambridge University Press.

11. Rodgers, W.H. 1994. *Environmental Law*. 2nd ed. West Publishing. The late Professor Rodgers taught law at the University of Washington for many years and is acknowledged as a leading authority on environmental law.

12. See the U.S. Supreme Court's decision: *Kleppe v. Sierra Club*, 427 U.S. 390, 96 S.Ct. 2718 (1976). We will explore Common Law more thoroughly in chapter 2.

13. Forum shopping is the practice in which a party to a lawsuit sues in the court or jurisdiction that is most likely to produce a decision that is favorable to them. This can only happen when more than one court has jurisdiction over the dispute, as is often true in NEPA cases. Court rules and legal cases strongly discourage forum shopping.

14. 42 U.S.C. §4321.

15. 42 U.S.C. §4331(a).

16. 42 U.S.C. §4331(b).

17. 42 U.S.C. §4331(c).

18. See Anderson, F. R., Glicksman, R. L., Mandelker, D. R., & Tarlock, A. D. 1999. *Environmental Protection: Law and Policy* (Law School Casebook Series). 3rd ed. Little, Brown, at p. 195. See also Calvert Cliffs' Coordinating Committee, *Inc. v. U.S. Atomic Energy Commission*, 449 F.2d 1109 (D.C. Cir. 1971).

19. See Keys, D. 2022. What's Substantive NEPA and Who Should Care? LinkedIn Articles. https://www.linkedin.com/pulse/whats-substantive-nepa-who-should-care-david-keys

20. *Strycker's Bay Neighborhood Council v. Karlen* (1980), 444 U.S. 223, 227–228, 100 S. Ct. 497. See Bogg, J.P. 1993. Procedural vs. Substantive in the NEPA Law: Cutting the Gordian Knot. *Environmental Professional, 15*(1), 25–33.

21. 42 U.S.C. § 4332(2)(C).

22. NEPA regulations are found at 40 CFR Parts 1500–1508.

23. 40 CFR §1508.18.

24. 40 CFR §1508.27.

25. See USEPA. National Environmental Policy Act Review Process. https://www.epa.gov/nepa/national-environmental-policy-act-review-process.

26. 40 CFR §1508.1(d).

27. Alternatives as required by §102(2)(E) of NEPA, 42 U.S.C. §4332(2)(E).

28. National Preservation Institute. Environmental Assessment. https://www.npi.org/environmental-assessment.

29. Middleton, T. 2021. What is an Environmental Impact Statement? American Bar Association, Teaching Legal Docs, March 2. https://www.americanbar.org/groups/public_education/publications/teaching-legal-docs/teaching-legal-docs--what-is-an-environmental-impact-statement-/.

30. USEPA. NEPA. Environmental Impact Statement (EIS) Database. https://cdxapps.epa.gov/cdx-enepa-II/public/action/eis/search.

31. See Pepper, E. 2015. Never Eliminate Public Advice: NEPA Success Stories. Natural Resources Defense Council, February 1. https://www.nrdc.org/resources/never-eliminate-public-advice-nepa-success-stories.

32. See Mackenzie, A., & Ruiz, S. 2023. No, NEPA Really Is a Problem for Clean Energy. Four Reasons the Arguments of NEPA Defenders Miss the Mark. Institute for Progress. Infrastructure, August 17. https://ifp.org/no-nepa-really-is-a-problem-for-clean-energy/.

33. Middleton, T. 2021. What is an Environmental Impact Statement? American Bar Association, Teaching Legal Docs, March 2. https://www.americanbar.org/groups/public_education/publications/teaching-legal-docs/teaching-legal-docs--what-is-an-environmental-impact-statement-/.

34. See the review in Berry, J.F., & Dennison, M.S. 2000. *The Environmental Law and Compliance Handbook*. McGraw-Hill.

35. 42 U.S.C. §4332(2)(D).

36. 40 CFR §1508.25.

37. *Society for Protection of New Hampshire Forests v. Brinegar*, 381 F. Supp. 282 (D.N.H. 1974).

38. Exec. Order No. 13,807, 82 Fed. Reg. 40,463 (Aug. 15, 2017), revoked by Exec. Order No. 13,990, 86 Fed. Reg. 7,037 (Jan. 20, 2021).

39. 40 CFR § 1501.9(e)(3) (2020).

40. See the comprehensive review in: Woolverton, M.B. 2021. Turning a Blind Eye to the Environment: Why Eliminating the Cumulative Effects Analysis Is Not Entitled to Chevron Deference. *Alabama Law Review, 73,* 199–219.

41. Administrative Procedures Act. 5 U.S.C. §701–706. Limitations under the APA are discussed in Birnbach, I. 1997. Newly Imposed Limitations on Citizens' Right to Sue for Standing in a Procedural Rights Case. *Fordham Environmental Law Journal, 9*(2), 311–336.

42. See, especially: Mandelker, D.R. 2023. *NEPA Law & Litigation, 2d.* 2023 ed. Thomson Reuters/West Group.

43. See Berry, J.F., & Dennison, M.S. 2000. *The Environmental Law and Compliance Handbook*. McGraw-Hill.

44. *Vermont Yankee Nuclear Power Corp. v. Natural Resources Defense Council*, 435 U.S. 519, 98 S.Ct. 1197 (1978).

45. *Robertson v. Methow Valley Citizens Council*, 490 U.S. 332, 109 S.Ct 1835 (1989).

46. See Daniels, B., Follett, A. P., & Salzman, J. 2020. Reconsidering NEPA. *Indiana Law Journal, 96*(3), 865–910. See also Frank, S.X. 2020. Is NEPA Still the Best Model for Environmental Protection? A Case for the NEPC. *Georgetown Environmental Law Review,* November 8. https://www.law.georgetown.edu/environmental-law-review/blog/is-nepa-still-the-best-model-for-environmental-protection-a-case-for-the-nepc/.

47. Earthjustice. 2023. The People's Environmental Law: National Environmental Policy Act, January 9. https://earthjustice.org/feature/national-environmental-policy-act

48. 42 U.S.C. § 4332(2)(F).

49. Executive Order 12114 (1979).

50. Eccleston, C.H. 2008. *NEPA and Environmental Planning: Tools, Techniques, and Approaches for Practitioners*. CRC Press.

51. *Lucas v. South Carolina Coastal Council*, 505 U.S. 1003, 12 S. Ct. 2886 (1992).

Chapter 2

Environmental Laws and Regulations

Chapter Learning Outcomes

At the end of this chapter, the reader will be able to:

- *Identify the three branches of government in the United States and describe their functions.*
- *Distinguish between a law and a regulation.*
- *List the types of common-law actions.*
- *Describe five major environmental laws in the United States.*
- *Identify the major international agreements that protect aspects of the environment.*

We will examine many environmental laws and regulations and other legal mechanisms throughout this book. Before we begin, however, let's define some important legal terms. In a typical lawsuit, there are two parties involved: the *plaintiff* is the person bringing the suit, while the *defendant* is the person against whom the lawsuit is brought. These parties can be individual people, companies, or even government entities. The plaintiffs in a lawsuit are pursuing a specific *remedy* if they win, which can be money, or to force the defendant to stop or change an action (an injunction).

Before we begin our examination of the ways the legal system protects and manages the environment, it is important that we establish exactly what we mean by ***laws*** and ***regulations***.

2.1 BRANCHES OF GOVERNMENT, AND THE ORIGINS OF LAWS AND REGULATIONS

In the United States of America (and many other countries), the government consists of three branches, with the expectation that each branch will function to balance excesses by the others. The three branches are the legislative branch, the executive branch, and the judicial branch. Our discussion will focus on the federal or national government, but it is important to remember that each state has a government that parallels the federal system (most states are essentially identical to the federal system). We will examine the function of each branch.

Legislative Branch

The legislative branch consists of the Senate and the House of Representatives. The familiar task of the legislative branch is to pass laws. Typically, a bill is introduced in either the Senate or the House. The bill is then assigned to a committee that may hold hearings, conduct investigations or studies, and ultimately issue a report to the Senate or House with a recommendation on approval. If the bill passes a vote, it is then sent to the other house (either House of Representatives or Senate) for its consideration. If different versions of the same bill are approved by each house, then a joint committee will attempt to reach a compromise that satisfies both houses. If members of both houses pass the bill, it becomes an Act. The Act becomes a Law if the president signs it (or fails to veto it within ten days). Once an Act has become a Law, it is *codified* within the United States Code (usually abbreviated "U.S.C.") and is often considered a *statute*.[1]

Executive Branch

The executive branch of government consists of the president and the Cabinet. The president is elected by the people, and the members of the Cabinet are nominated by the president and approved by the Senate. Examples of Cabinet-level positions of greatest importance in environmental laws are the Administrator of the U.S. Environmental Protection Agency and the Secretary of the Department of the Interior. In general, the function of the executive branch is to "execute," or administer, the laws. Most environmental laws contain a provision in which the legislature "delegates" to the executive branch the authority to conduct the day-to-day administration of the laws by way of "regulations." If Congress has failed to delegate the authority to create regulations, then the president may do so via an "Executive Order."[2] We will see many examples of environmental laws, regulations, and executive orders throughout this book.

Judicial Branch

The judicial branch is the third branch of government in the United States, composed of the federal District Courts, the Circuit Courts of Appeal, and the U.S. Supreme Court. The purposes of the federal courts are, in general, to make decisions regarding disputes among parties over issues that are national in scope.[3] In deciding the cases that are brought before them, judges interpret the relevant statutes and regulations, apply constitutional principles, and often follow legal precedents.

The lowest courts, or trial courts, in the federal system are the geographically delineated District Courts. These are the first courts to which a plaintiff turns to ask that a decision be made.[4] The District Courts hear arguments and examine evidence, often with the assistance of a jury, then render their decision based on the record established during the trial.

The loser in the District Court may then appeal to a regional Circuit Court of Appeals. There are eleven regional Circuit Courts of Appeals, plus the Court of Appeals for the Federal Circuit in Washington D.C. that hears appeals only in a few types of cases. It is important to note that very few cases are automatically appealed as a matter of right. Instead, the Circuit Courts of Appeals pick the cases they will hear on appeal based on their importance. If the Circuit Court of Appeals denies an appeal, that is usually the end of the case for the parties. If the Circuit Court does agree to hear an appeal, that appeal is limited to the record established in the District Court (that is, no new evidence is allowed).

The loser in the Circuit Court may then appeal to the U.S. Supreme Court. Given the large number of cases appealed to the Supreme Court every year, only a small percentage are heard.[5] Once again, if an appeal is heard, it is limited to the record established in the original District Court, and it is usually only the attorneys representing the parties that are invited to speak to the Supreme Court. There is no opportunity for appeal within the judicial system beyond the U.S. Supreme Court. Given the small number of environmental cases heard by the Supreme Court each year, it is not surprising that they are of considerable importance.

We will examine many laws and regulations, and many instances when the legal system is an important part of managing, regulating, and protecting the environment. As we proceed, it is important to remember how our legal system operates.

Courts of Law and Courts of Equity

For many years, there were two types of courts in the United States and elsewhere, called "courts of law" and "courts of equity." While the concepts are still with us today, the courts generally are not.

The origins of law and equity courts began with the early years of what became known as English Common Law. In the early years (twelfth and thirteenth centuries), the common law terminology was followed exactly (hence, "word for word"). These early law courts could issue punishments for those who transgressed, and issue damages and other remedies.

But through the years, the Church of England saw that there were circumstances where a court of law could not supply an adequate remedy to a disagreement. So, the Church created its own courts to solve these non-law issues, with judges called "chancellors" who created equitable remedies to problems. In these equity courts (also known as "chancellory courts" or "chancery courts"), the chancellor has the power to order acts rather than damages. As a result, injunctions, specific performance of an act, and vacatur are remedies available in equity. We will discuss a few of these remedies below.

Over time, these courts evolved into the courts of law (which resolved issues of law) and the courts of equity (which resolved equitable issues). Courts of law and equity came to the United States along with other English Common Law in the eighteenth century but have gradually merged into a single court system in nearly all jurisdictions. Remnants of equity law are still with us, however.[6] What we know as modern "family law" is mostly equity law. Equitable issues are quite common in environmental law too, and we will see numerous examples.

2.2 HISTORICAL ORIGINS OF ENVIRONMENTAL LAW: COMMON-LAW ACTIONS

In addition to the federal and state environmental laws that we will discuss throughout this book, it is important to examine the various common-law theories that may be asserted as a basis for environmental liability claims. Common law, in contrast to statutory law, refers to those general principles and rules pertaining to persons and property that have developed over time from custom and usage, and from court decisions that have ruled on those customs and usages. Common-law actions in the United States were based on similar laws that originated in twelfth-century England, although U.S. common law has matured and followed its own path. Common-law claims are an important part of the history of U.S. environmental law,

A useful way of conceptualizing the distinction between common law and the statutory law we discussed above is to think of the common law simply as non-statutory law. Statutory law pertains to all federal and state laws specifically created by legislative enactment. Common law, on the other hand, refers to all remaining rules, principles, and customs not derived from express legislative authority, but rather from the decisions of judges.

Under modern circumstances, legal claims for environmental injury to persons and property may fall under a particular environmental statute, such as liability for a polluter under the Comprehensive Environmental Response, Compensation, and Liability Act (CERCLA).[7] Nevertheless, common-law claims are an important part of environmental law today, so we will examine some important common-law actions.

Common-law liability may stem from personal injuries and property damage caused by environmental conditions under common-law theories of negligence, nuisance, trespass, and strict liability for abnormally dangerous activities. For example, an individual or company may be sued by environmental authorities to clean up contaminated property, and neighboring property owners may assert statutory and common-law claims for damage caused to their properties. Except for negligence, common-law liability for environmental conditions is imposed *without regard to fault.*

Trespass

Trespass is a common-law action to remedy an unauthorized invasion of a person's right to exclusive possession of the land. By contrast, an actionable invasion of a person's right to the use and enjoyment of land is a nuisance. For example, one court equated knowledge on the part of a defendant concerning the release of sludge from its industrial waste dump onto the plaintiff's land with an intentional invasion of the landowner's property interest.[8] Since the defendants had been notified that their sludge was entering the plaintiffs' mine and the defendants did nothing to stop it, the plaintiffs had a valid common-law action for trespass.

Trespass is a *strict liability* claim. This means that plaintiffs in trespass actions may succeed even in the absence of proof of actual harm or injury.[9] In an influential statement of common law, Section 158 of the Restatement (Second) of Torts states that "one is subject to liability to another for trespass, irrespective of whether he thereby causes harm."[10]

Nuisance

Nuisance represents another form of common-law action for an unreasonable interference with the use and enjoyment of property. The law of nuisance recognizes two conflicting rights: (1) property owners have a right to control their land and use it to benefit their best interests, and (2) the public and neighboring landowners have a right to prevent unreasonable use that substantially impairs the peaceful use and enjoyment of other land. Nuisance law is based on the principle that "[o]ne must use his own property so that his neighbor's comfortable and reasonable use and enjoyment of his estate will not be unreasonably interfered with or disturbed."[11]

The legal remedy for a nuisance may be in the form of monetary damages and/or injunctive relief. In determining whether an activity constitutes a nuisance, the courts usually weigh the utility and public acceptability of the activity with the extent of harm or cost of compensation for the injury to the complaining party's property.[12] A court may choose to deny injunctions when the resulting losses to the polluters would greatly exceed the benefits to their victims.[13]

There are two types of nuisance actions: *public nuisance* and *private nuisance*. Although the two categories often overlap, there is an important distinction between them. Public nuisance actions arise when there has been an unreasonable interference with a property right "common to the general public," such as when an improperly managed public landfill violates the public's right to a safe environment.[14] Private nuisance actions, on the other hand, involve unreasonable interferences with an individual's use and enjoyment of his or her property, such as when a neighboring landowner discharges hazardous substances onto the property of the adjacent landowner.[15] In either case, the claim of nuisance requires that the pollution-causing activity be unreasonable. The unreasonable use element of nuisance balances the rights of adjoining property owners.[16]

For a private person to recover damages for a public nuisance, the individual must show that the injury suffered was in some way different from that suffered by other members of the general public. A private party generally has no right of action for a public nuisance, because "[i]t is the province of the public authorities to procure redress for public wrongs."[17]

A *private nuisance* is the unreasonable, unusual, or unnatural use of one's property so that it substantially impairs the right of another to peacefully enjoy his property. It is anything that disturbs the free use of one's property, or which renders its ordinary use or physical occupation uncomfortable. Section 822 of the Restatement (Second) of Torts defines the elements of a private nuisance as follows: "One is subject to liability for a private nuisance if, but only if, his conduct is a legal cause of an invasion of another's interest in the private use and enjoyment of land, and the invasion is either (a) intentional and unreasonable, or (b) unintentional and otherwise actionable under the rules controlling liability for negligent or reckless conduct, or for abnormally dangerous conditions or activities."[18]

In addition, a nuisance can be characterized as either *temporary* or *permanent*, which can be a key factor in determining the proper measure of damages for the injury inflicted on the plaintiff's property. As a general rule, a nuisance is considered temporary if it may be abated, and permanent if abatement is impracticable or impossible.

Strict Liability for Ultrahazardous Activities

Under the common-law theory of strict liability for ultrahazardous or abnormally dangerous activities, liability is imposed without regard to fault for injuries caused by an ultrahazardous activity. This longstanding theory has been applied in the modern context of injuries caused by toxic, hazardous, and radioactive substances.

A useful characterization of this common-law action is in Section 519 of the Restatement (Second) of Torts, which establishes the elements of strict liability for harm caused by abnormally dangerous activities as follows:

1. One who carries on an abnormally dangerous activity is subject to liability for harm to the person, land, or chattels of another resulting from the activity, although he has exercised the utmost care to prevent the harm.
2. This strict liability is limited to the kind of harm, the possibility of which makes the activity abnormally dangerous.[19]

Although the vast majority of cases involving injury to a person or property from hazardous substances are based on violations of various federal and state statutes and regulations, the common-law cause of action for ultrahazardous or abnormally dangerous activities may be applicable in certain environmental situations, such as *toxic torts.* [20]

Negligence

Negligence is usually a component of another common-law action, but it also appears as its own separate action in environmental cases. Negligence occurs when someone fails to use such care as a reasonably prudent person would use to avoid causing foreseeable harm to someone else. This common-law doctrine can be used in the context of environmental liability when someone negligently releases a hazardous substance into the environment that causes personal injury or property damage to others. Negligence includes conduct falling below the standard of a reasonably prudent person under similar circumstances. As stated in Section 282 of the Restatement (Second) of Torts, negligence encompasses all conduct "which falls below the standard established by law for the protection of others against unreasonable risk of harm."

The duty of care component of negligence is dependent on proof that it was reasonably foreseeable that a particular act or omission by the defendant would cause injury to the plaintiff. No duty of care can be imposed on a defendant to take precautions against events that could not be reasonably foreseen. In determining whether the plaintiff's injury from a hazardous substance was reasonably foreseeable by the defendant, courts will generally employ a standard of reasonable care. Under this standard, the courts will consider what a reasonably prudent person should have done or the applicable standard of care for a particular industry.

Environmental Toxic Torts

The field of *environmental toxic torts* has become an important discipline within the field of environmental law. An environmental toxic tort is a claim for damages arising from exposure to a harmful chemical or substance. Toxic tort claims usually arise when a defendant's negligence causes a plaintiff to sustain personal injuries as a result of exposure to a toxic substance. Common law claims of nuisance, trespass, and strict liability for abnormally dangerous activities are usually asserted for the recovery of damage to property. Personal injury damages may be sought under these theories, in addition to property damage claims. However, recovery of personal injury-related damages is generally based on negligence or another common-law theory.

Today, environmental toxic torts are mostly related to injuries or potential injuries from a wide variety of substances, including pesticides, heavy metals, benzene, polychlorinated biphenyls (PCBs), and other contaminants in the air, water, or soil. A plaintiff in a toxic tort action must prove that the substance to which she or he was exposed is toxic. She or he must prove that there was sufficient exposure to cause harm, such as ingestion of a sufficient dosage and concentration of the substance to bring about a harmful effect. The plaintiff must also prove that exposure to the substance was caused by the defendant's wrongful conduct and that the exposure was not from a substance that occurs naturally in the environment.

The Public Trust Doctrine

The public trust doctrine is a legal principle establishing that certain natural and cultural resources are preserved by the State for use by the public. Natural resources held in public trust can include navigable waters, wildlife, or land. The public is considered to be the owner of the resources, and the government protects and maintains these resources for the public's use.[21]

The public trust doctrine is often used today in the context of water bodies. Most lakes, rivers, and streams in the United States are maintained under the public trust doctrine, usually for the purposes of drinking water and recreation.

The public trust doctrine is a common source of litigation when citizens are concerned that public resources are being used for private purposes.

2.3 MODERN ENVIRONMENTAL LAWS AND REGULATIONS

In the twenty-first century, environmental laws (statutes) and administrative regulations are far more important than common-law actions. We will encounter many examples of environmental laws and regulations in subsequent chapters. Here is a short list of some of the major federal environmental laws we will discuss:[22]

National Environmental Policy Act (NEPA)

The passage of NEPA resulted in the creation of the Environmental Protection Agency. The "Congressional declaration of purpose" section at the beginning of the Act explains that NEPA's purpose is, "To declare a national policy which will encourage productive and enjoyable harmony between man and his environment; to promote efforts which will prevent or eliminate damage to the environment and biosphere and stimulate the health and welfare of man; to enrich the understanding of the ecological systems and natural resources important to the Nation; and to establish a Council on Environmental Quality." While the EPA is heavily involved with NEPA, the White House also appoints the Council on Environmental Quality, which oversees the implementation of NEPA across all agencies.[23]

Clean Air Act (CAA)

According to the "Congressional findings and declaration of purpose" of the CAA, the Act was designed to "protect and enhance the quality of the Nation's air resources" and to promote research and provide assistance to state and local governments in an effort to combat air pollution. The CAA is overseen by the EPA.[24]

Clean Water Act (CWA)

The CWA's "Congressional declaration of goals and policy" states that the "objective of this chapter is to restore and maintain the chemical, physical, and biological integrity of the Nation's waters." The act provides for research, enforcement, and state assistance in efforts to curb water pollution. The CWA is overseen by the EPA.[25]

Resource Conservation and Recovery Act (RCRA)

RCRA was passed to regulate solid waste in the United States. In the "Congressional findings" section, it is further provided that due to the passage of other environmental laws and regulations (such as the CAA and CWA, more solid waste was being created, and there needed to be more regulation on hazardous waste disposal. The RCRA is overseen by the EPA.[26]

Comprehensive Environmental Response, Compensation, and Liability Act (CERCLA)

CERCLA, also known as Superfund, is a law that placed a tax on certain businesses in industries engaged in work with hazardous materials. The purpose of the tax was to provide funding to clean up any hazardous materials disposal sites if those businesses no longer existed. CERCLA is overseen by the EPA.[27]

Endangered Species Act (ESA)

The ESA governs the conservation of fauna in the United States. The "Policy" section of the ESA states that it is "the policy of Congress that all Federal departments and agencies shall seek to conserve endangered species and threatened species." The act further provides that members from various agencies are to be placed on a task force, but the final determination of animals that satisfy listing as endangered species is made by the Secretary of the Interior or the Secretary of Commerce. The ESA is overseen by the U.S. Fish and Wildlife Service and the National Oceanic and Atmospheric Administration.[28]

Marine Protection, Research and Sanctuaries Act (MPRSA)

The MPRSA was passed to regulate dumping in the oceans. The "Congressional finding, policy, and declaration of purpose" section provides "it is the policy of the United States to regulate the dumping of all types of materials into ocean waters." The MPRSA further provides a framework for the regulation of dumping and enforcement of those regulations. The MPRSA is overseen by the EPA.[29]

2.4 ENVIRONMENTAL LITIGATION

We will encounter many challenges to environmental laws and regulations in later chapters. An exhaustive review of environmental legal challenges is beyond the scope of this book, but there are many extensive works available that do just that.[30]

Environmental litigation (lawsuits) tends to follow two general paths. First, there are lawsuits between private parties. These can be between individuals, organizations, or corporations and are often based on the common-law principles we discussed above. There are a few environmental laws, like CERCLA, that encourage the parties involved with pollution to work out relative liability through lawsuits between the parties. We will discuss CERCLA and relative liability in chapter 9.

Most environmental litigation, however, involves federal agencies. In one kind of case, the state or federal government sues an industry or industrial source for not complying with a particular environmental law. For example, the U.S. EPA could prosecute a power plant for violating the terms of its permit to discharge pollutants into the air in violation of the CAA. The remedy may include civil or criminal fines, or even criminal prison terms.

In an alternative kind of case, an industry or source sues the government, arguing that the enforcement agency exceeded its authority in making regulations or in applying them in the particular case. If the industry prevails, then its regulatory burden may be reduced or regulations might be changed.

Environmental litigation can involve challenges or defenses to specific state or federal laws or regulations, challenges to constitutional provisions, or government actions. We will discuss specific forms of litigation under a variety of situations later in the book.[31]

2.5 INTERNATIONAL ENVIRONMENTAL LAW

International environmental law is the least familiar aspect of environmental law to most people, including environmental lawyers. One reason is that it may not seem clear what, exactly, environmental law is. We will use Professor Guruswamy's definition that international environmental law "consists of international law dealing with the environment as found, primarily, in international agreements (also called treaties, conventions, or pacts), together with the international mechanisms for implementing them; and secondarily, in international customary law (the common law of the international community)."[32] International law consisting of international agreements is also known as *hard international law*, and international customary law is known as *soft international law*.

We will encounter many examples of international environmental laws in subsequent chapters.

Enforcement

As with any legal system, an important part of international environmental law is the mechanism of enforcement. No law can be effective without a mechanism to enforce compliance by the regulated community. Enforcement of hard and soft law differs in important ways.

Hard international law is the law of bilateral and multilateral agreements such as treaties, charters, conventions, protocols, and so on that create specific legally binding obligations on two or more signatory nations. These agreements use a variety of specific regulatory tools for designating responsibilities and authorizing enforcement. Enforcement can be dispute resolution mechanisms specified in the agreement, various financial charges, or even trade limitations. While these mechanisms are regarded as hard law, there are many examples of situations where these mechanisms have failed, leading to the argument that hard law does not always work as intended.[33]

By contrast, customary international law, or soft law, is probably the weakest form of international environmental law. It is often based on little more than customary trade practices and habits, and is largely unenforceable. Soft laws are usually expressed as political or value statements, rather than legally binding provisions of an agreement. Agreements of this type exist only as long as the parties involved continue to support them.

Examples of Bilateral and Multilateral Agreements

We will encounter many examples in this book of environmental agreements between two or more governments, but here are a few examples to help us understand how they work.

A good starting point is the Convention on International Trade in Endangered Species, known as CITES. CITES was adopted in 1963 at a meeting of members of the IUCN (The World Conservation Union). As of 2023, 184 countries, including the United States, have become members. Under the CITES agreement, all member nations "shall not allow trade in specimens of species included in Appendices I, II and III except in accordance with the provisions of the present Convention."[34] The idea is to provide an international mechanism to protect endangered or threatened plants and animals from exploitation. Enforcement is controversial and depends on the legal apparatus of the countries involved. We will discuss CITES further in chapter 4.

The *Paris Agreement*, also known as the *Paris Accords* or the *Paris Climate Accords*, is an international treaty on climate change. The agreement, adopted in 2015, covers climate change mitigation, adaptation, and finance. The Paris Agreement was negotiated by 196 parties at the 2015 United Nations Climate Change Conference near Paris. As of February 2023, 195 members of the United Nations Framework Convention on Climate Change (UNFCCC) are parties to the agreement. The United States withdrew from the agreement in 2020 but rejoined in 2021. Under the agreement, each country must determine, plan, and regularly report on its contributions. Mechanisms for enforcement are left intentionally unclear. We will discuss climate change in chapter 12.

We will discuss the concept of biological diversity in chapter 3, but we can think of it as the variety of living things in a geographical area. There is general worldwide agreement that biological diversity is under threat by human actions. In response, the Convention on Biological Diversity (CBD) was created in 1992. The objectives of the CBD are the conservation of biological diversity, the sustainable use of its components, and the fair and equitable sharing of the benefits arising out of the utilization of genetic resources, including by appropriate access to genetic resources and by appropriate transfer of relevant technologies, taking into account all rights over those resources and to technologies, and by appropriate funding. There is a clear attempt to ensure that the countries that supply the biological and genetic diversity to the rest of the Conference will profit. Once again, however, enforcement has been controversial and incomplete.

The *Montreal Protocol* is an international treaty designed to protect the stratospheric ozone layer by phasing out the production of numerous substances (like chlorofluorocarbons, or CFCs) that are responsible for ozone depletion. It entered into force in 1989 and has been ratified by almost 200 countries. It has undergone nine revisions. The provisions of the Protocol include the requirement that the Parties to the Protocol base their future decisions on the current scientific, environmental, technical, and economic information, that is, assessed through panels drawn from the worldwide expert communities. The Montreal Protocol is essentially voluntary, so enforcement is difficult. However, due to its widespread adoption and implementation, it has been hailed as an example of successful international cooperation. In fact, there is considerable scientific evidence that stratospheric ozone depletion has slowed, and that ozone levels will return to normal by the 2040s.

There are many other international agreements that deal with environmental issues of many kinds. We will explore some of these as we continue our discussions.

* * *

Review Questions

1. List the three branches of the U.S. government, and describe the duties of each.
2. What is a statute, and who produces it?
3. What is a regulation, who produces it, and what does it do?
4. How do common-law actions differ from statutory actions?
5. What is the difference between a private nuisance and a public nuisance?
6. What are environmental toxic torts?
7. Describe three environmental laws and their basic structure.

Case Analysis

In 1984, the U.S. Supreme Court issued its decision in the case *Chevron, U.S.A. v. Natural Resources Defense Council*.[35] The Chevron case established a precedent under the Administrative Procedures Act[36] called "Chevron deference" that instructed courts to defer to federal agencies when their authority is left ambiguous by a law (as it frequently is). Federal judges are seldom educated in the science that supports most federal agency decisions on the environment, and the Chevron case prevented judges from substituting their judgment for that of agency experts.

However, in June 2024, the Supreme Court released its decision in *Loper Bright Enterprises v. Raimondo*, which overruled the Chevron decision and removed the deference to federal agency expertise, replacing it with an earlier test that leaves the decisions in the hands of judges.

The *Loper* case involved a rule by the National Marine Fisheries Service, which is part of the National Oceanic and Atmospheric Administration of the Department of Congress. The rule stated that commercial herring fishermen must have an observer onboard to collect conservation data. The issue was either the observer would be paid by the agency (which was the fishermen's interpretation) or by the fishermen (the agency's view). The lower courts had deferred to the agency's interpretation as the Chevron case required, but the Supreme Court disagreed.

The following is part of the Court's opinion in the Loper decision. The fundamental question is this: If there is ambiguity in a regulation, should that ambiguity be resolved by the agency or by the Court under the Administrative Procedures Act (APA)? Read the Court's language and answer the questions below.

> *Chevron* cannot be reconciled with the APA, as the Government and the dissent contend, by presuming that statutory ambiguities are implicit delegations to agencies. As *Chevron* itself noted, ambiguities may result from an inability on the part of Congress to squarely answer the question at hand, or from a failure to even "consider the question" with the requisite precision. In neither case does an ambiguity necessarily reflect a congressional intent that an agency, as opposed to a court, resolve the resulting interpretive question. And many or perhaps most statutory ambiguities may be unintentional. As the Framers recognized, ambiguities will inevitably follow from "the complexity of objects, . . . the imperfection of the human faculties," and the simple fact that "no language is so copious as to supply words and phrases for every complex idea."
>
> Courts, after all, routinely confront statutory ambiguities in cases having nothing to do with *Chevron*—cases that do not involve agency interpretations or delegations of authority. Of course, when faced with a statutory ambiguity in such a case, the ambiguity is not a delegation to anybody, and a court is not

somehow relieved of its obligation to independently interpret the statute. Courts in that situation do not throw up their hands because "Congress's instructions have" supposedly "run out," leaving a statutory "gap." Courts instead understand that such statutes, no matter how impenetrable, do—in fact, must—have a single, best meaning. That is the whole point of having written statutes; "every statute's meaning is fixed at the time of enactment." So instead of declaring a particular party's reading "permissible" in such a case, courts use every tool at their disposal to determine the best reading of the statute and resolve the ambiguity.

Perhaps most fundamentally, *Chevron*'s presumption is misguided because agencies have no special competence in resolving statutory ambiguities. Courts do. The Framers, as noted, anticipated that courts would often confront statutory ambiguities and expected that courts would resolve them by exercising independent legal judgment. And even *Chevron* itself reaffirmed that "[t]he judiciary is the final authority on issues of statutory construction" and recognized that "in the absence of an administrative interpretation," it is "necessary" for a court to "impose its own construction on the statute." *Chevron* gravely erred, though, in concluding that the inquiry is fundamentally different just because an administrative interpretation is in play. The very point of the traditional tools of statutory construction—the tools courts use every day—is to resolve statutory ambiguities. That is no less true when the ambiguity is about the scope of an agency's own power—perhaps the occasion on which abdication in favor of the agency is *least* appropriate.[37]

Questions

1. Is the Court right when it says that Congress did not assume that agencies have greater expertise to resolve ambiguities? Why do you think this is the case?
2. The Court in *Loper* said, "*Chevron* gravely erred, though, in concluding that the inquiry is fundamentally different just because an administrative interpretation is in play." Does it matter that the interpretation involved a scientific issue decided by a scientific agency? Should it matter?
3. What are the possible changes in federal agency actions that might result from the *Loper* decision?
4. Did the Court in *Loper* make the right decision? Why, or why not?

NOTES

1. Once a law has been codified, it is still frequently known by its title and section number within the original Act. For example, the section of the Clean Water Act that applies to the filling of wetlands is usually called "Clean Water Act §404," rather than "33 U.S.C. §1344" [volume 33 of the U.S. Code, section 1344].

2. For example, in 1977, President Carter signed Executive Order 11,990 (42 Fed. Reg. 26,967 (1977)) authorizing the CEQ to promulgate regulations under NEPA because the statute contained no rule-making authority.

3. This includes disputes under most (but not all) federal environmental laws, but may also include disputes among parties located across state boundaries from each other.

4. The exceptions are the District Court for the District of Columbia and the U.S. Claims Court, which are not truly geographical.

5. In actual practice, the loser in the Circuit Court of Appeals petitions the Supreme Court for a "writ of certiorari." If you examine written court decisions, you may see the citation "cert. denied" before the Supreme Court reference, indicating that the Supreme Court did not grant certiorari, and the appeal was denied.

6. There are numerous scholarly discussions of this merger. Some examples: McCormick, C. 1928. The Fusion of Law and Equity in United States Courts. *N.C. Law Review, 6,* 283297; and Funk, K. 2022. Equity's Federalism. *Notre Dame Law Review, 97,* 2057–2093. https://scholarship.law.nd.edu/ndlr/vol97/iss5/9.

7. 42 U.S.C. §9601 et seq. (1980), also known as the "superfund" law. We will discuss CERCLA in chapter 9.

8. *Curry Coal Co. v. M.C. Arnoni Co.*, 266 A.2d 678 (Pa. 1970).

9. See, for example, *Ream v. Keen*, 112 Or.App. 197, 828 P.2d 1038, aff'd, 314 Or. 370 (1992) (under Oregon law, actual damage is unnecessary to establish liability for an intentional trespass).

10. Restatement (Second) of Torts § 158.

11. Patz v. Farmegg Products, Inc., 196 N.W.2d 557, 560 (Iowa 1972), quoting *Bates v. Quality Ready Mix Co.*, 261 Iowa 696, 702, 154 N.W.2d 852, 857 (1967).

12. See, for example, *Stevinson v. Deffenbaugh Indus.*, 870 S.W.2d 851 (Mo.App. 1993).

13. See, for example, *Pate v. City of Martin*, 614 S.W.2d 46 (Tenn. 1981).

14. See, for example, *Blair v. Anderson*, 570 N.E.2d 1337 (Ind.App. 1991).

15. See, for example, *O'Neal v. Department of the Army*, 852 F. Supp. 327 (M.D. Pa. 1994); Mel Foster Co. *Properties v. American Oil Co.*, 427 N.W.2d 171 (Iowa 1988).

16. See, for example, *Stevinson v. Deffenbaugh Indus.*, 870 S.W.2d 851 (Mo.App. 1993).

17. *Blair v. Anderson*, 570 N.E.2d 1337 (Ind.App. 1991).

18. Restatement (Second) of Torts § 822.

19. Restatement (Second) of Torts § 519.

20. Toxic torts are common-law actions in which there is personal injury or property damage due to exposure to toxic or hazardous substances.

21. For a comprehensive review of the public trust doctrine including its historical origins and applications, see: Ryan, E., Curry H., & Rule, H. 2021. Environmental Rights for the 21st Century: A Comprehensive Analysis of the Public Trust Doctrine and Rights of Nature Movement. *Cardozo Law Review, 42,* 2447–2576. https://ir.law.fsu.edu/articles/721.

22. Information from the Library of Congress at: https://guides.loc.gov/environmental-law/federal-laws. The U.S. EPA's website has a more complete list with descriptions of federal environmental laws and executive orders: https://www.epa.gov/laws-regulations/laws-and-executive-orders.

23. 42 U.S.C. §4321, et seq.

24. 42 U.S.C.§7401, et seq.

25. 33 U.S.C. §1251, et seq.

26. 42 U.S.C. §6901, et seq.

27. 42 U.S.C. §9601, et seq.

28. 16 U.S.C. §1531.

29. 33 U.S.C. §1401.

30. See, for example: Brown, K. A., & Hogan, A. M. (Eds.). 2019. *Environmental Litigation: Law and Strategy*. American Bar Association; and Hill, B.E. 2022. *Environmental Justice: Legal Theory and Practice.* 5th ed. ELI Press; among many others.

31. Columbia Law School's Sabin Center for Climate Change Law maintains an excellent web page that organizes environmental lawsuits according to categories. https://climatecasechart.com/us-climate-change-litigation/.

32. Guruswamy, L. 1995. International Environmental Law: Boundaries, Landmarks, and Realities. *Natural Resources & Environment, 10*(2), 43–45. 76–77.

33. See, for example, the review by Kilduff, T. 2019. The Difficulties of Enforcing Global Environmental Law. *Georgetown Environmental Law Review*, February 1. https://www.law.georgetown.edu/environmental-law-review/blog/214/.

34. CITES Article II(4).

35. 467 U.S. 837, 104 S.Ct. 2778.

36. Administrative Procedures Act, 5 U.S.C. §701–706. See the discussion of the APA in chapter 1.

37. *Loper Bright Enterprises v. Raimondo*, 603 U.S. ____ (2024).

Chapter 3

The Modern Approach to Ecosystems

Chapter Learning Outcomes

At the end of this chapter, the reader will be able to:

1. *Describe the characteristics of the three spheres of the earth.*
2. *Contrast the components of an organism, species, population, community, and ecosystem.*
3. *Distinguish between the biotic and abiotic components of an ecosystem.*
4. *Describe energy flow and trophic levels in a community.*
5. *Contrast the flow of materials in the major biogeochemical cycles.*
6. *Identify ways that federal laws can protect ecosystems.*

Now that we have learned something about the environmental legal system, let's turn our attention to environmental science. Specifically, we will examine some ecological principles that will help us understand the relationships between the various living and nonliving parts of the environment. We will do this through an investigation of the ecological principle of ecosystems, but first, a little background scientific information is in order.

3.1 THE SPHERES OF THE EARTH

It is useful to think of the earth as a large system that consists of four interdependent subsystems, or *spheres* (figure 3.1). These are the land, water, living things, and air. These four spheres are known as the *geosphere* (land), *hydrosphere* (water), *atmosphere* (air), and *biosphere* (living things). Their components are exchanged and cycled constantly to form dynamic and fully interactive systems. Still, it is useful to explore each sphere separately.

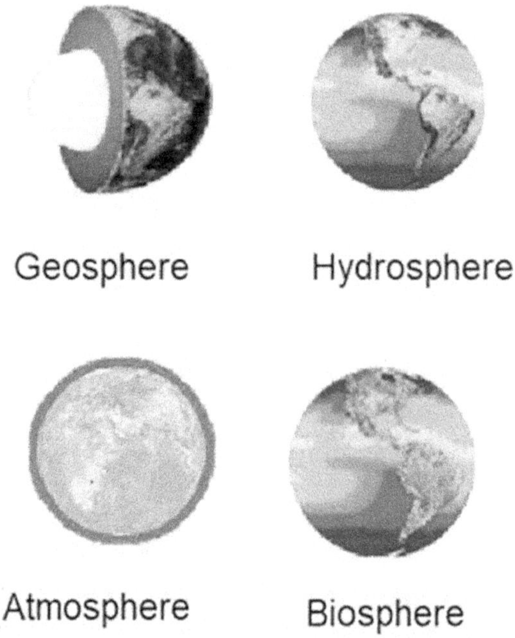

Figure 3.1 The Four Interdependent Subsystems, or *Spheres* of the Earth. *Source*: Images from NOAA: https://www.google.com/url?sa=t&rct=j&q=&esrc=s&source=web&cd=&ved=2ahUKEwjp7KKqkYyAAxWCTTABHeJnDCIQFnoECBYQAQ&url=https%3A%2F%2Fgml.noaa.gov%2Foutreach%2Finfo_activities%2Fpdfs%2FTBI_earth_spheres.pdf&usg=AOvVaw0dEpWak8oVKCacOiw9L-8y&opi=89978449

Geosphere

The geosphere consists of the continental and oceanic crust and all layers of the earth's interior. This includes all rocks, sediments, and soils, surface landforms, and the processes that shape the earth's surface, as well as other materials that make up the earth's crust (surface). It also includes the semi-solid materials beneath the crust and the liquid materials near the center of the earth. The surface of the geosphere is uneven, with mountain ranges like the Rockies and Andes, huge plains or flat areas like those in the American Midwest and Brazil, and deep valleys and trenches like those along the ocean floor.

The surface of the earth's geosphere is composed of the crust and the lithospheric mantle (or mantle lithosphere), the uppermost part of the mantle that is not subject to convection (movement due to heat). The *lithosphere* is underlain by the *asthenosphere,* which is the weaker, hotter, and deeper part of the upper mantle that experiences convection. Below that is the molten core of the earth, which is also subject to convection.

Hydrosphere

The hydrosphere consists of the liquid, solid, and gaseous water of the planet. Ranging from 10 to 20 kilometers in thickness, the hydrosphere extends from the earth's surface down several kilometers into the lithosphere and upward about 12 kilometers into the atmosphere.

Ninety-seven percent of earth's water is salty. The salty water collects in deep valleys along earth's surface that we call oceans. Water near the poles is much colder than water near the equator. The differences in temperature cause water to change physical states. Extremely low temperatures, like those found at the poles, cause water to freeze into a solid as a polar icecap, a glacier, or an iceberg. Very high temperatures, like those found at the equator, cause water to evaporate into a gas.

Only about 3 percent of the water in the hydrosphere is fresh water (non-salty). Fresh water falls as precipitation from the atmosphere down to earth's surface, as rivers and streams along the earth's surface (surface water), and as groundwater beneath earth's surface. Nearly 70 percent of the earth's fresh water, however, is frozen as ice sheets and glaciers. The part of the hydrosphere that is frozen is known as the *cryosphere*.

Atmosphere

The *atmosphere* consists of the layers of gases that surround the earth, which we call "air." It extends from chambers below the earth's surface to more than 10,000 kilometers above the surface. The upper portion of the atmosphere protects the organisms of the biosphere from the sun's ultraviolet radiation. It also absorbs and emits heat. Changes in temperature, moisture, and other factors are responsible for what we perceive as short-term weather and long-term climate. We will discuss the atmosphere, weather, and climate more fully in chapter 6.

Biosphere

The *biosphere* consists of all living things on the planet, which includes all of the microorganisms, plants, and animals of Earth. Within the biosphere, living things form *ecosystems* based on the physical characteristics and environment of a particular area. As we will discuss below, these ecosystems are arranged into *biomes*, and the biomes into the biosphere. Obviously, the components of the biosphere overlap with the other spheres. It's easy to imagine a flying insect that breathes and flies in the atmosphere, feeds on the geosphere, and reproduces in the hydrosphere.

3.2 SPECIES, POPULATIONS, AND ECOSYSTEMS

We have seen that the biosphere encompasses all living things on the planet. Let's examine the components of the biosphere, beginning with an individual living thing or an *organism*. Individual organisms are, of course, composed of even smaller components, including organs and organelles, elementary particles (electrons, neutrons, and protons), atoms, molecules, and compounds.

Determining what is or is not an organism is more difficult than you might think. For example, a dead animal is not considered to be an organism because it is no longer a living thing. But what do we do with viruses? A virus is composed primarily of a DNA or RNA strand, which may include a protein coat. However, viruses are usually not considered to be alive since they cannot reproduce without a host cell and do not have an independent metabolism like living things.[1] The question of whether viruses qualify as organisms is a matter of controversy and is undecided at this time. For our purposes, we will consider viruses to be parts of ecosystems, whether they are alive or not.

If we begin our organization with individual organisms, we can then arrange them into gradually larger, more inclusive categories of the biosphere in what we can call the levels of ecological organization (figure 3.2).

Organism

An individual living thing that has an organized structure can react to stimuli, grow, reproduce, adapt to its environment, and maintain a metabolism. You are an organism, but so is an oak tree or an individual bacterium. Organisms can be single-celled or multicellular. Organisms are generally members of a *species* that consists of genetically similar individuals with the potential to interbreed and form populations.

Population

A group of individuals of the same species in an area is a *population*. Some populations are very large, like some birds or fishes that may consist of many thousands of individuals, or small, like some predator populations that may consist of only a few hundred individual organisms.

Community

A group of populations of living things in an area is a *community*. Communities consist of all plant, animal, and microbe populations. They are usually named for the dominant plant type (e.g., Pine Forest or *Spartina* Marsh).

Figure 3.2 Levels of Ecological Organization. Living things in the environment are organized according to a specific pattern. *Source*: Images from NOAA coral reef ecosystem: https://www.noaa.gov/education/resource-collections/marine-life/coral-reef-ecosystems

Ecosystem

The living (biotic) community and the nonliving (abiotic) components of an environment constitute an *ecosystem*. The abiotic components may include water, minerals, weather, and so on. Because ecosystems incorporate both the living and nonliving parts of an environment, they are very useful as a focus for environmental laws and regulations, as we will see.

Biome

Large regions of the earth with similar ecosystems are known as *biomes*. Biomes can be either terrestrial or aquatic. The major biomes are Tropical Forest (Rainforest), Temperate Forest, Desert, Tundra, Taiga (Boreal Forest), Grassland, Savanna, Freshwater, and Marine.[2]

Biosphere

The part of the earth that contains all ecosystems.

Our focus in this book will be on ecosystems. Although the term has entered the language of environmental law only recently, the concept has been with us for centuries. A similar idea, one that is much more familiar to legal language, is the concept of ***habitat***. The U.S. Environmental Protection Agency defines an organism's habitat as "all the physical, chemical and biological attributes that affect or sustain the organisms within an ecosystem."[3] In other words, an organism's habitat is everything that it requires for its home. For example, the habitat of an easter gray squirrel (*Sciurus carolinensis*) is large, forested areas with plentiful nut trees (like oak, hickory, and walnut) as a food source, trees with nesting cavities for reproduction and security, and a permanent source of nutrients and fresh water. The term *habitat* is an important part of the laws that relate to the protection of threatened and endangered species, as we will discover shortly.

Another important biological issue is the question of how "big" a habitat should be. It depends on the species, of course. For some large, free-ranging species (e.g., wolves), the critical habitat may be many square miles. For smaller, stationary species (e.g., some plants), the habitat may be a few square meters. For migratory species, habitat shifts during the season. In designing habitat *reserves* for protected species, it is often necessary to protect connecting *wildlife corridors* for movement between *patches* of habitat in order to create a habitat that is large enough for the target species (figure 3.3).

In this same sense, biologists often refer to the ***ecological niche*** of an organism or a population. The ecological niche can be defined as "the position of a species within an ecosystem, comprising both the habitat requirements and

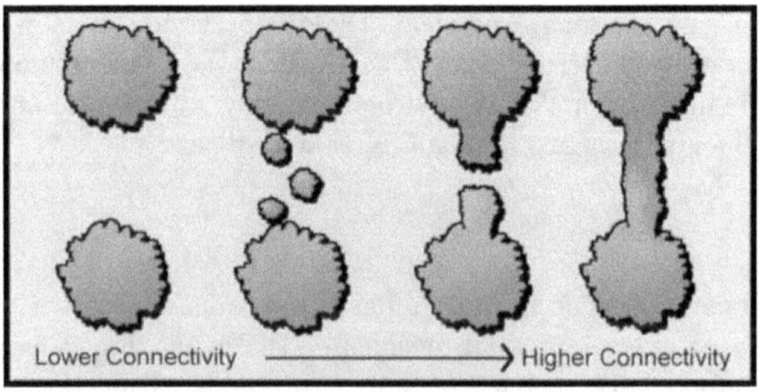

Figure 3.3 Wildlife Corridors. Connecting patches with corridors can benefit biodiversity by providing access to other areas of habitat, increasing gene flow and population viability, enabling recolonization of patches, and providing habitat. *Source*: USDA National Agroforestry Center. https://www.fs.usda.gov/nac/buffers/guidelines/2_biodiversity/3.html

the functional role of a species."[4] We can think of an organism's ecological niche as its job description. For example, there may be an ecological niche in a forest for a large seed-eating bird with a powerful bill. This is a niche that might be filled by a cardinal in a forest in the eastern United States, or a parrot in South America, or a grosbeak in Asia. Under proper circumstances, the niche might be filled by a very different animal, like a flying mammal or an insect. Ecological studies of niches have a very rich history.

Over the past several decades, conservation biologists have emphasized the importance of protecting ***biodiversity***,[5] which we define as the number of different species of organisms (plants and animals) in an ecosystem. Healthy ecosystems consist of numerous groups of interactive and interdependent plants and animals. Biologists have determined that global (and often local) biodiversity is rapidly declining, which is a matter of great urgency for several reasons. First, rapidly increasing human populations are degrading the natural environment at an alarming rate. Second, science is finding new uses for biodiversity to relieve human suffering (e.g., new medicines) and abate environmental destruction. Third, biodiversity is lost through increasing rates of extinction caused by destruction of natural habitats.

In response to losses of biodiversity in the United States, the Department of the Interior launched a National Biological Survey (NBS) in 1993 under the auspices of the U.S. Fish and Wildlife Service (USFWS) to assess the condition of biodiversity in the United States, and to recommend measures to reverse the trend in lost biodiversity. The NBS has demonstrated a series of successes, even though it has dealt with funding issues and political difficulties.

3.3 ENERGY FLOW AND TROPHIC LEVELS

All living organisms require energy to grow, reproduce, and move. Energy cannot be created from nothing, so it must be transferred through ecosystems. The primary source of energy for almost every ecosystem on earth is the sun.[6] **Primary producers**, also known as **autotrophs**,[7] are plants that use energy from the sun to produce glucose for food through the process of photosynthesis. These primary producers are then consumed by various levels of **consumers**, or **heterotrophs**. **Primary consumers**, or herbivores, are in turn eaten by secondary consumers and so on, so that energy flows from one trophic level, or level of the food chain, to the next. The easiest way to demonstrate this energy flow is with a food chain. In figure 3.4, each link in the chain represents a trophic level, and the arrows show energy being passed along the chain. At the beginning of a food chain is always the primary producer. In terrestrial ecosystems, most primary producers are plants, and in marine ecosystems, most primary

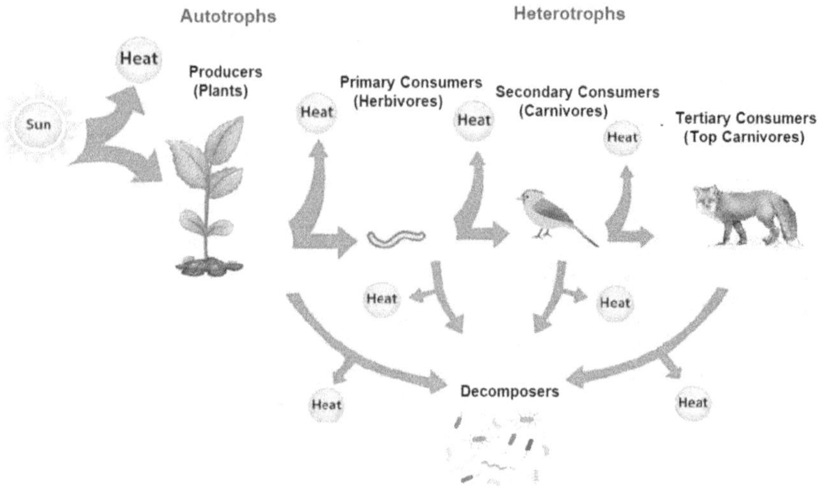

Figure 3.4 A Simple Food Chain and Energy Flow in a Terrestrial Ecosystem. *Source*: Modified from Innoclazz. Own work. Wikimedia Commons. https://commons.wikimedia.org/wiki/File:Energy_flow_through_various_levels_of_ecosystem.jpg

producers are phytoplankton. Both produce most of the nutrients and energy needed to support the rest of the food chain in their respective ecosystems.

All the **biomass**[8] generated by primary producers is called **gross primary productivity**. Everything that remains after the primary producers have used the energy they require for respiration is **net primary productivity**. This is the portion that is available to be consumed by the primary consumers and passed along the food chain. In terrestrial ecosystems, primary productivity is highest in warm, wet places with plenty of sunlight, like coastal marshes and tropical forest regions. Deserts generally have the lowest primary productivity.

A simplified way to show relationships between organisms in an ecosystem is by using a **food chain**, which shows the flow of energy from one set of organisms to another (figure 3.4). Since energy conversion is never perfect, wasted energy at each step is given off as heat.

It is also useful to look at food chain relationships as **trophic pyramids** to show the flow of energy through ecosystems. Each step of the pyramid represents a trophic level, beginning with primary producers at the bottom. As shown in figure 3.5, the width of each step represents the rate of energy flow through each trophic level. The steps get smaller farther up the pyramid because some of that energy is changed to a form that cannot be consumed by organisms at the next higher step in the food chain (usually as heat). This happens at every step in the pyramid.

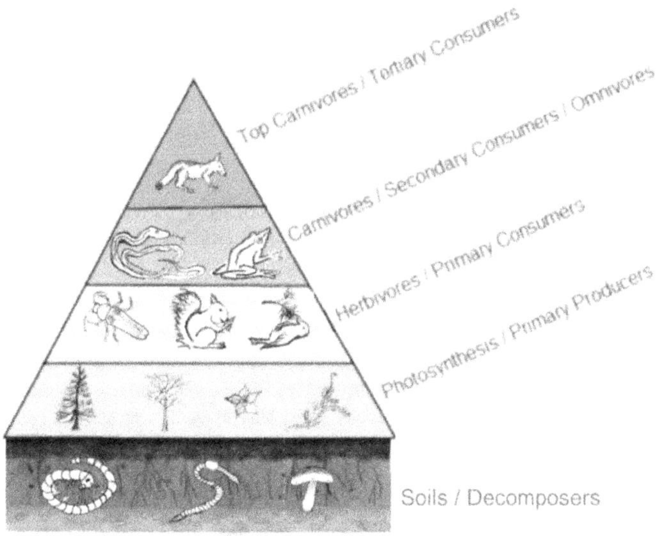

Figure 3.5 A Trophic Pyramid. Thompsma. *Source*: Own Work. Wikimedia Commons. https://commons.wikimedia.org/wiki/File:Trophiclevels.jpg

Not all of the energy generated or consumed in one trophic level will be available to the organisms in the next higher trophic level. At each level, much of the biomass (and energy) is changed to heat during respiration and is unavailable for consumption. Much of the rest is excreted as waste, while some plants and animals die without being eaten, so their biomass is not passed on to the next trophic level. The waste and dead matter are then broken down by decomposers and the nutrients are recycled into the soil to be taken up again by plants, but most of the energy is changed to heat and is lost to the environment (figure 3.4). On average, only about 10 percent of energy stored as biomass in a trophic level is passed from one level to the next. This is known as *the ten percent rule* and it limits the number of trophic levels that can be supported by an ecosystem.[9]

Examining food chains is a useful way to visualize relationships among organisms and trophic levels, but food chains are an oversimplification. In the real world, there are many interactions among groups of organisms. A *food web* is a representation of the interactions between the food chains in an ecosystem that represents the flow of nutrients and energy within the ecosystem. It is also known as a consumer-resource system. Food webs provide a useful overview of the interactions among organisms with each other and their environment. A simple aquatic food web is in figure 3.6.

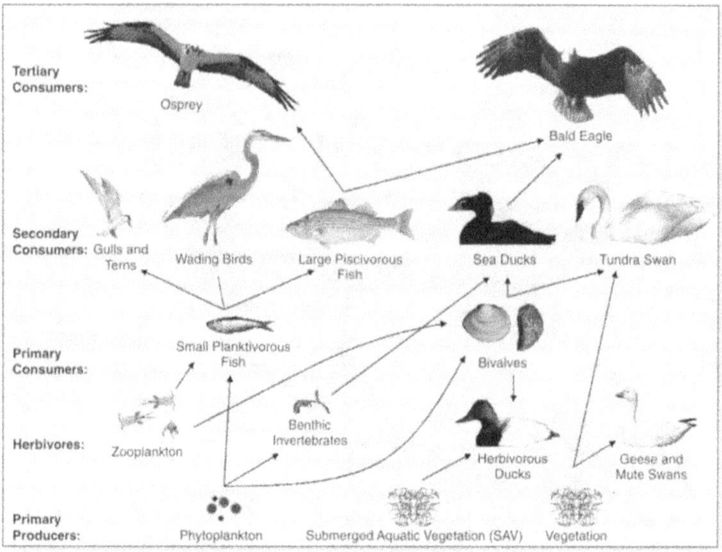

Figure 3.6 A Food Web of an Aquatic Ecosystem. USGS. *Source*: Chesapeake Bay Waterbird Food Web Illustration from Circular 1316. https://www.usgs.gov/media/images/chesapeake-bay-waterbird-food-web-illustration-circular-1316

3.4 BIOGEOCHEMICAL CYCLES

Ecosystems consist of both the living (biotic) and nonliving (abiotic) components of the environment. These components exist in constantly shifting patterns of cycles that are called *biogeochemical cycles*. Let's examine the most environmentally significant of these cycles as we explore the concept of ecosystems.

Biogeochemical cycles describe the ways in which elements (like nitrogen or carbon) or compounds (such as water or carbon dioxide) move between the biotic and abiotic components of an ecosystem. These elements and compounds move constantly between the biosphere, geosphere, hydrosphere, and atmosphere in ways that support ecosystems.

The biogeochemical cycles of greatest importance to living organisms include the water, carbon, nitrogen, phosphorus, and sulfur cycles. An exploration of these biogeochemical cycles will give us greater insight into the functioning of ecosystems.

The Water Cycle

The *water cycle*, or *hydrologic cycle*, is the most familiar biogeochemical cycle to most people. The water cycle stores and distributes the earth's fixed water supply, as shown in figure 3.7. The water cycle is a global cycle because there is a large reservoir of water in the atmosphere as well as in the hydrosphere, especially the oceans. Water is of tremendous significance to life on the earth.[10] In fact, biologists recognize that life as we know it cannot exist without water, due to its three main properties. First, water acts as a solvent for most solids (more than other solvents), which provides a mechanism for the movement of nutrients and wastes between and within cells. In addition, water is a major medium for transporting nutrients within and between ecosystems.

Second, water molecules are polar (different electrical charges at either end), which causes water molecules to attract each other and permits water to remain a liquid throughout a relatively large range of temperatures (between freezing and boiling). Third, water has the unusual property of having greater density as a liquid than as a solid. This is why ice floats on water and why it expands when it freezes. Liquid water in cracks between rocks freezes and shatters the rocks, which is responsible for the rapid weathering and erosion of the earth's crust.

The energy that powers the water cycle comes from the sun and involves three major processes—evaporation, precipitation, and transpiration. Solar

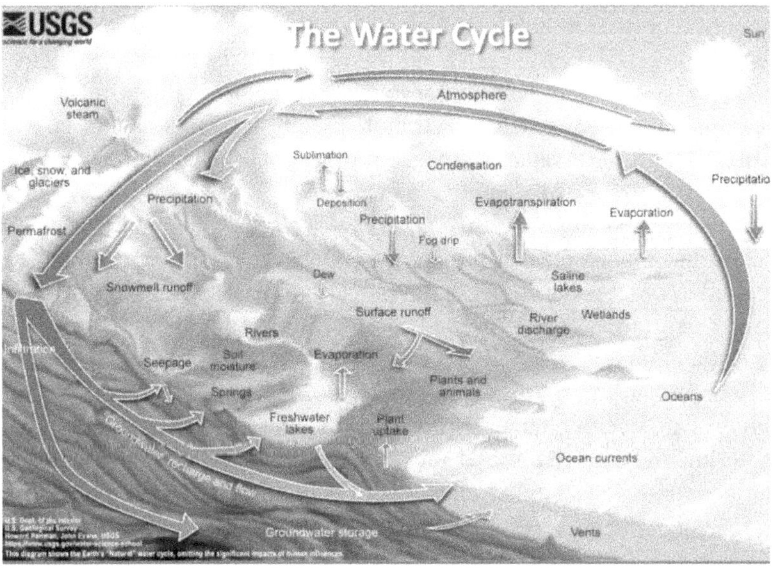

Figure 3.7 The Water Cycle. *Source*: U.S. Geological Survey. https://www.usgs.gov/media/images/natural-water-cycle-jpg

energy causes the *evaporation* of water from water bodies and soils. Evaporation converts liquid water into water vapor in the atmosphere, where it rises and condenses into liquid water. Gravity then draws the water back to the earth's surface as *precipitation* (rain, snow, sleet, and dew). Most water vapor (about 85 percent) in the atmosphere comes from the oceans, with the rest coming from land. About 90 percent of the water that reaches the atmosphere from land comes from the surfaces of plants through a process called *evapotranspiration*, or simply *transpiration*, which drives the internal circulation of the plant.

Water falling as precipitation takes various forms. Most precipitation falling on terrestrial ecosystems becomes *surface runoff.* Much of this water flows into streams and lakes, which eventually carry water back to the oceans, from which it can evaporate to repeat the cycle. Some surface water leaches into the upper layer of soils, and some evaporates from soils, lakes, and streams back into the atmosphere. Some precipitation is converted to ice that may be stored as *glaciers*. Some precipitation sinks through soil and permeable rock formations to underground layers of sand, gravel, and porous rock called *aquifers,* where it is stored as *groundwater.*

Only about 0.1 percent of the earth's water ends up in the biotic (living) components of ecosystems. Roots of plants absorb some of this water, most of which evaporates from plant leaves back into the atmosphere through the process of transpiration. Some water combines with carbon dioxide during *photosynthesis* to produce sugars and other carbohydrates. Eventually, these compounds are broken down in plant cells, which release water back into the environment.

Water is an important contributor to the constant recycling of materials on the earth's surface. Surface water runoff replenishes streams and lakes and contributes to the erosion and movement of soil and rock fragments. Water is the primary sculptor of the earth's landscape.

The water cycle also serves to purify water by removing materials such as nutrients and toxins. During the processes of evaporation and precipitation, water undergoes a natural distillation process that removes impurities. Water flowing as surface water or groundwater is filtered and partially purified by chemical and biological processes, such as bacteria.

It is important to note that only about 0.024 percent of the earth's water supply is available to humans as liquid freshwater in accessible groundwater deposits and in lakes, rivers, and streams. The rest is too salty or polluted with impurities for us to use, is stored as ice, or is too deep underground to extract economically with existing technologies.

The Carbon Cycle

Carbon is the basic building block of all organic molecules, including proteins, carbohydrates, fats, and nucleic acids (DNA and RNA) that are necessary for life. Carbon circulates through the biosphere, geosphere, atmosphere, and parts of the hydrosphere, in the *carbon cycle* shown in figure 3.8. It is an important abiotic component of ecosystems because it is necessary for respiration and photosynthesis by the earth's living organisms.

Most of the earth's carbon is stored in rocks and sediments. The rest is located in the ocean, atmosphere, and in living organisms. These are the reservoirs through which carbon cycles. Atmospheric carbon is in the form of carbon dioxide (CO_2) gas, which makes up 0.036 percent of the volume of the atmosphere and is also dissolved in water. Carbon dioxide is an important greenhouse gas, as we will discuss further in chapters 6 and 12. As a result, if the carbon cycle removes too much CO_2 from the atmosphere, the atmosphere will cool, and if it generates too much, the atmosphere will get warmer. Even slight changes in this cycle, caused by natural or human factors, can affect

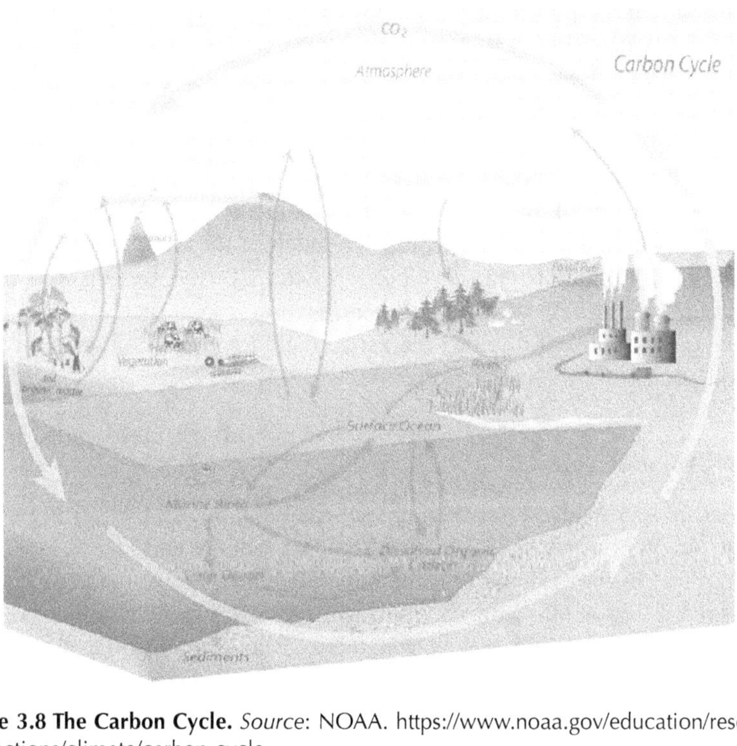

Figure 3.8 The Carbon Cycle. *Source*: NOAA. https://www.noaa.gov/education/resource-collections/climate/carbon-cycle

weather and climate, and ultimately help determine the types of life and ecosystems that can exist on the earth.

We can think of carbon in the carbon cycle as moving from one storage reservoir to another through various mechanisms. For example, in a food chain, plants move carbon from the atmosphere into the biosphere through photosynthesis. Plants use energy from the sun to combine carbon dioxide chemically with hydrogen and oxygen from water to create sugar molecules. Animals that eat plants digest the sugar molecules to get energy for their bodies. Respiration, excretion, and decomposition release the carbon back into the atmosphere or soil, continuing the cycle.

The earth's oceans play a critical role in carbon storage since they hold about fifty times more carbon than the atmosphere. Two-way carbon exchange can occur between the oceans' surface waters and the atmosphere, but carbon may be stored for centuries at the deepest depths of the oceans.

Rocks like limestone and fossil fuels like coal and oil are reservoirs that store carbon from plants and animals that lived millions of years ago. When these organisms died, slow geologic processes trapped their carbon and transformed it into the forms we see today. Processes such as erosion and volcanic activity release this carbon back into the atmosphere. Burning fossil fuels by humans in cars or power plants releases carbon back into the atmospheric reservoir very quickly.

As we will explore in later chapters, human activities have a huge impact on the carbon cycle. Burning fossil fuels, changing land use, forest fires, and using limestone to make concrete all transfer significant quantities of carbon into the atmosphere. As a result, the amount of carbon dioxide in the atmosphere is rapidly rising; it is already greater than at any time in the last 3.6 million years. The ocean absorbs much of the carbon dioxide that is released from burning fossil fuels, but this extra CO_2 is lowering the ocean's pH through a process called *ocean acidification*. Ocean acidification interferes with the ability of marine organisms (including corals, crustaceans, and snails) to construct shells and skeletons.

The Nitrogen Cycle

Nitrogen is a critical nutrient for the survival and success of all organisms and the health of ecosystems. Around 78 percent of the earth's atmosphere is composed of nitrogen, making it the most abundant atmospheric gas by far. Atmospheric nitrogen occurs as dinitrogen gas (N_2), which cannot be used directly by living organisms such as plants and can limit nitrogen availability in ecosystems.[11] The nitrogen cycle is a key component in many ecosystem processes such as decomposition and primary production, and nitrogen availability can alter the rate of these processes. Nitrogen has several forms including dinitrogen gas (N_2), nitrogen oxide (NO), nitrogen dioxide (NO_2), ammonia (NH_3), ammonium ions (NH_4^+), and ammonium nitrate (NH_4NO_3).

Through a series of processes, nitrogen is converted by microbial action through fixation, assimilation, ammonification, nitrification, and denitrification. These processes make up the *nitrogen cycle* and play an important role for all living organisms on the earth (figure 3.9).

Nitrogen fixation is the process by which nitrogen gas (N_2) is transformed into ammonium (NH_4^+), a form of nitrogen that can be used by plants. Through this process, nitrogen moves from the atmosphere into the soil, where plants can absorb it through their root systems. A small percentage of fixation can occur via abiotic activities such as lightning, but the majority of nitrogen fixation occurs naturally in soils by bacteria that have a symbiotic relationship with plants. In exchange for energy from photosynthesis, the bacteria will fix nitrogen into a usable form for the plant by using the enzyme *nitrogenase*. Nitrogen fixation by bacteria can also produce forms of nitrogen that can be utilized by various organisms.

Nitrogen-fixing bacteria, which live as symbionts on plant roots, fix the most nitrogen. These bacteria can invade the roots of legume plants and provide the plants with usable nitrogen to assist with protein production and energy in the form of carbon for the symbiont bacteria. This process is

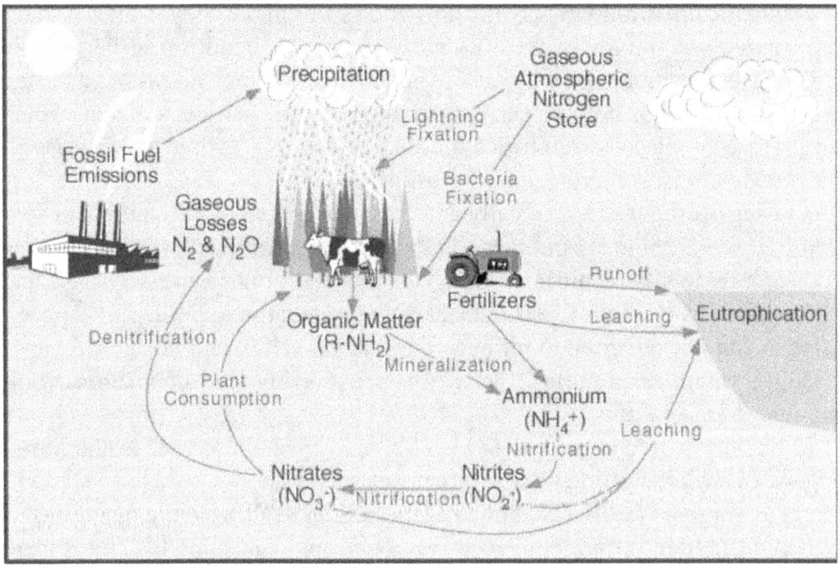

Figure 3.9 The Nitrogen Cycle. *Source*: Nitrogen cycle NOAA Climate Program Office. https://cpo.noaa.gov/Divisions-Programs/Earth-System-Science-and-Modeling/Atmospheric-Chemistry-Carbon-Cycle-Climate-AC4/Improved-understanding-of-nitrogen-cycle

beneficial to agriculture as leguminous plants can assist with returning nitrogen into the soil to promote plant growth.

Assimilation of inorganic nitrogen is the process by which organic nitrogen compounds form from inorganic nitrogen compounds in an ecosystem. Plants use these ions to make proteins and nucleic acids. The assimilation process occurs when nitrates enter a cell and are reduced to ammonia. This ammonia is incorporated into organic compounds for the rest of the cycle.

Soil nitrogen can be derived from dead organic materials. *Ammonification* or mineralization is the process where bacteria incorporate nitrogen into amino acids and release the excess nitrogen as ammonium ions (NH_4^+) into the soil. These ammonium ions are then readily available for uptake by plants for protein synthesis and microorganisms that require it for growth.

Nitrification is a two-part microbial process in which ammonium ions are converted into nitrates and nitrites. This process provides extra available nitrogen for plants to take in via their roots. Through the process of nitrification, ammonium, produced by *ammonification*, found in soils is transformed into nitrites (NO_2^-) and nitrates (NO_3^-). Nitrates are used by plants and plant-consuming animals and are formed by ammonia-oxidizing bacteria. Nitrites are not readily available to plants and animals but can be converted to nitrates by bacteria. Nitrate is the form most living plants use to absorb nitrogen.

Denitrification follows nitrification and is the process by which nitrates are returned to the atmosphere as nitrogen gas by denitrifying bacteria in soils. Denitrification generally occurs in anoxic environments with exhausted oxygen levels. This process can lead to a loss in soil nitrogen content, which needs to be replaced. Denitrification can also occur during the process of harvesting crops, soil erosion, and burning.

The nitrogen cycle is critically important in maintaining natural ecosystems. However, humans intervene in the nitrogen cycle in several ways. We add large amounts of nitric oxide (NO) into the atmosphere when N_2 and O_2 combine as we burn fossil fuels at high temperatures. In the atmosphere, this gas can be converted to nitrogen dioxide gas (NO_2) and nitric acid vapor (HNO_3), which can return to the earth as damaging acid deposition, commonly called acid rain.

In addition, we add nitrous oxide (N_2O), a greenhouse gas, to the atmosphere through the action of anaerobic bacteria on livestock wastes and commercial inorganic fertilizers applied to the soil. We release large quantities of nitrogen stored in soils and plants as gaseous compounds into the atmosphere through the destruction of forests, grasslands, and wetlands. We upset the nitrogen cycle in aquatic ecosystems by adding excess nitrates to bodies of water through agricultural runoff and discharges from municipal sewage systems.

According to the 2005 Millennium Ecosystem Assessment,[12] since 1950, human activities have more than doubled the annual release of nitrogen from the land into the rest of the environment. Most of this is from the greatly increased use of inorganic fertilizer to grow crops. The amount released is projected to double again by 2050. Excessive input of nitrogen into the air and water contributes to pollution, acid rain, and other problems to be discussed in later chapters.

The Phosphorus Cycle

Phosphorus circulates through water, the earth's crust, and living organisms in the *phosphorus cycle* (figure 3.10). In contrast to the cycles of water, carbon, and nitrogen, the phosphorus cycle does not include the atmosphere.

Phosphorus is an essential nutrient for living processes. It is an important component of nucleic acids and phospholipids in cell structure. As calcium phosphate, it makes up the supportive components of our bones. Phosphorus is often the limiting nutrient in aquatic ecosystems.

Phosphorus occurs in nature primarily as the phosphate ion (PO_4^{3-}). Phosphates enter surface water runoff as a result of human activity or as natural surface runoff that occurs when it is leached from phosphate-containing rock

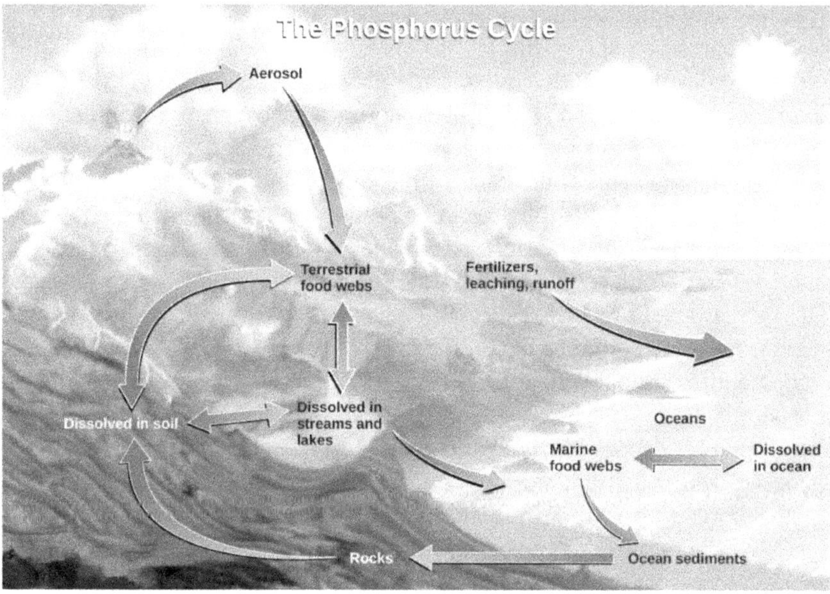

Figure 3.10 The Phosphorus Cycle. *Source*: Image modified from CNX OpenStax Biology. Modification of work by John M. Evans and Howard Perlman, USGS. This file is licensed under the Creative Commons Attribution 4.0 International license.

by weathering, thus sending phosphates into rivers, lakes, and the ocean. Phosphate-containing ocean sediments form primarily from the bodies of ocean organisms and from their excretions. However, in remote regions, volcanic ash, aerosols, and mineral dust may also be significant sources of phosphate. This sediment is moved to land over geologic time by the uplifting of areas of the earth's surface.

The major reservoir for phosphorus is phosphate salts containing phosphate ions (PO_4^{3-}) in terrestrial rock formations and ocean bottom sediments. The phosphorus cycle is slow compared to the water, carbon, and nitrogen cycles.

As water runs over exposed rocks containing phosphorus, it slowly erodes away compounds that contain phosphate ions. The dissolved phosphate is absorbed by the roots of plants and by other producers. Phosphorus is transferred by food webs from such producers to consumers, eventually including detritus feeders and decomposers. In both producers and consumers, phosphorus is a component of biologically important molecules such as nucleic acids and energy transfer molecules such as ADP and ATP. It is also a major component of vertebrate bones and teeth. Phosphate can be lost from the cycle for long periods when it washes from the land into streams and rivers and is carried to the ocean. There it can be deposited as marine sediment and remain trapped for millions of years. At some point in the future, geological processes may uplift and expose these deposits, phosphate can be eroded, and the cycle can start over again.

Terrestrial soils typically contain little phosphate, so it is often the limiting factor for plant growth on land. If phosphorus (as phosphate salts mined from the earth) is applied to the soil as an inorganic fertilizer, then plant growth may be increased. This is the basis of most agricultural crop fertilization.

Phosphorus also limits the growth of producer populations in many freshwater streams and lakes because phosphate salts are only slightly soluble in water. When phosphate is added to water as fertilizer runoff, it may dramatically increase plant growth, especially algae.

Anthropogenic activities have affected the phosphorus cycle by removing large quantities of phosphate from the earth to make fertilizer and reducing phosphorus in tropical soils by clearing forests. Soil that is eroded from fertilized crop fields carries large quantities of phosphates into streams, lakes, and the ocean, where it stimulates the growth of producers such as algae.

Phosphorus-rich runoff from the land can produce huge populations of algae, which can upset chemical cycling and other processes in lakes, leading to a process called *eutrophication*. Eutrophication occurs when a body of water receives an excessive nutrient load, particularly phosphorus and nitrogen, which results in an overgrowth of algae (also known as *phytoplankton*). As the algae die and decompose, oxygen is depleted from the water. The loss of oxygen in the water causes the death of aquatic organisms, such as fish and invertebrates. We will discuss these concepts further in chapter 7.

The Sulfur Cycle

Sulfur is another essential element for life because it is an important constituent of many proteins and other biologically important substances. In addition, sulfur compounds can be used as chemical oxidants or reductants in respiration and other cell processes. The global sulfur cycle involves the transformation of sulfur species through different oxidation states, which play an important role in both geological and biological processes.

Sulfur circulates through the biosphere in the sulfur cycle, as shown in figure 3.11. Much of the earth's sulfur is stored underground in rocks and minerals as elemental sulfur, or sulfate salts (SO_4^{2-}) buried under sediments.

Sulfur enters the atmosphere from several natural sources. Hydrogen sulfide (H_2S) is a colorless, poisonous gas with a smell like rotten eggs. Hydrogen sulfide gas is released from active volcanoes and from organic matter that is broken down by anaerobic bacteria in flooded terrestrial ecosystems like wetlands.

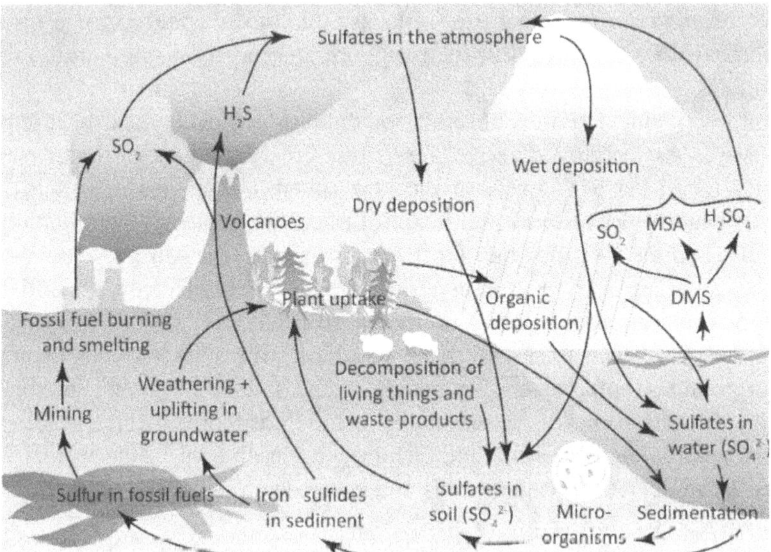

Figure 3.11 The Sulfur Cycle. *Source*: Lappalainen, Hanna K. et al. (2016) Pan-Eurasian Experiment (PEEX): toward a holistic understanding of the feedbacks and interactions in the land–atmosphere–ocean–society continuum in the northern Eurasian region, Atmospheric Chemistry and Physics, 16(22):14421-14461. doi:10.5194/acp-16-14421-2016

Sulfur dioxide (SO_2) is a colorless, suffocating gas that comes from natural sources like volcanoes and from industrial processes that involve the burning of fossil fuels. Sulfur dioxide is one of the criteria pollutants that is regulated under the federal Clean Air Act, which will be discussed in chapter 6.

In terrestrial ecosystems, sulfur is deposited in four major ways: precipitation, direct fallout from the atmosphere, rock weathering, and geothermal vents. Atmospheric sulfur is found in the form of sulfur dioxide (SO_2), and as rain falls through the atmosphere, sulfur dioxide reacts with water vapor to form weak sulfuric acid (H_2SO_4). This is known as acid rain, which will be discussed in chapter 6.

Sulfur can also fall directly from the atmosphere in a process called *fallout*. In addition, the weathering of sulfur-containing rocks releases sulfur into the soil. These rocks originate from ocean sediments that are moved to land by the geologic uplifting of ocean sediments. Terrestrial organisms can then make use of these soil sulfates (SO_4^-) in metabolic processes and release the sulfur back into the atmosphere as hydrogen sulfide (H_2S) gas when the organisms die and decompose.

Sulfur enters the ocean via runoff from land, from atmospheric fallout, and from underwater geothermal vents. Some ecosystems rely on organisms called *chemoautotrophs* that use sulfur as a biological energy source instead of photosynthesis. This sulfur then supports marine ecosystems in the form of sulfates.

Anthropogenic activities have altered the nitrogen cycle in several important ways. As mentioned above, fossil fuel combustion in the industry has greatly increased SO_2 levels in the atmosphere, which leads to acid rain and damage to ecosystems and health. In addition, extensive cultivation of legumes and the use of certain fertilizers has altered the sulfur cycle.

The use of nitrogen fixation in agriculture has greatly increased worldwide food production but has led to an increase in the emission of nitrogen into the atmosphere. This excess nitrogen can drain from soils into water sources underground or enter water systems via runoff. Excess nitrogen can contribute to eutrophication (discussed above in the Phosphorus Cycle), leading to issues such as algae blooms due to nitrogen enrichment in the water. This process can decrease oxygen levels and cause death in aquatic ecosystems.

3.5 LAWS REGARDING SPECIES, POPULATIONS, AND ECOSYSTEMS

There are few references to ecosystems in federal laws. Most major environmental laws were originally written in the 1970s, at a time before ecosystems were part of the national environmental dialogue. Because these laws change only slowly, references to ecosystems have rarely appeared so far.

An example is the requirement that "critical habitat" be designated for all endangered species under Section 4 of the federal ESA.[13] Critical habitat is defined in ESA §3 as "the specific areas within the geographical area occupied by the species . . . on which are found those physical or biological features . . . essential to the conservation of the species."[14] The use of the term "critical habitat" here means much the same thing as an ecosystem. In fact, the designation of critical habitat for a particular species offers protection for the endangered species in question, but also all of the other species (biotic factors) and the abiotic factors in the environment.

An example of how this works is the case of the northern spotted owl (*Strix occidentalis*). The spotted owl inhabits old-growth forests in the northwestern United States, but these forests were under pressure from the logging industry. In the 1980s, conservationists proposed federal protection for the northern spotted owl, which, in turn would protect the old-growth forest ecosystem and its inhabitants from logging. The idea of a single, important species is familiar to ecosystem ecologists as the concept of a ***keystone species*** (a species that exerts a large, stabilizing influence throughout an ecosystem, despite its relatively small numerical abundance). Thus, the federal ESAhas the effect of protecting ecosystems, even though the word is not used in the statute.

The designation of the northern spotted owl was challenged by forestry interests, and the case went back and forth through the courts for years. The U.S. Supreme Court supported the designation in favor of the spotted owl in 1992.[15] The story did not end there, though. In the last few days of the Trump Administration in 2021, the U.S. Fish and Wildlife Service removed about one-third of the owls' critical habitat from federal protection. This move was met by lawsuits from environmental groups, and the issue is ongoing.

There are other federal laws that use the same indirect protection of ecosystems that is used by the ESA. An example is the Marine Mammal Protection Act, which protects critical habitat "including the rookeries, mating grounds, and areas of similar significance for each species of marine mammal."[16] We will discuss these and other related laws in chapter 4.

By contrast, some more recent federal laws have specifically mentioned ecosystems and ecosystem protection. An example is the Infrastructure Investment and Jobs Act (IIJA)[17] of 2021, also known as the Bipartisan Infrastructure Law. Title VIII of the IIJA is titled *Natural Resources-Related Infrastructure, Wildfire Management, and Ecosystem Restoration*, and is designed in part to invest millions of dollars annually to restore habitat connectivity for aquatic species around the country and advance habitat restoration, invasive species control, conservation of at-risk and listed species, and benefits to several significant ecosystems.

While federal laws have been slow to adopt the terminology of ecosystems, federal administrative agencies tell a very different story. In fact, some of the most intensive uses of the term "ecosystem" are by federal agencies with strong environmental responsibilities. These agencies include the U.S. Environmental Protection Agency (USEPA),[18] the National Oceanic and Atmospheric Administration (NOAA),[19] the U.S. Fish and Wildlife Service (USFWS),[20] the U.S. Forest Service (U.S. Department of Agriculture, USDA),[21] and the National Park Service,[22] among others.

Some state laws have embraced the concept. California passed its Natural Community Conservation Planning Act (NCCPA)[23] which specifically protects natural communities and ecosystems. We will discuss NCCPs in chapter 4.

Other states have created agencies and regulations that support ecosystems. For example, the state of Florida has an Office of Ecosystem Projects[24] that administers Florida's laws and regulations designed to protect ecosystems.

* * *

Review Questions

1. Contrast the structures of the earth that make up the geosphere, hydrosphere, atmosphere, and biosphere.
2. Describe the components of an organism, species, population, community, and ecosystem.
3. What are the abiotic and biotic components of an ecosystem?
4. Describe the roles of producers, primary consumers, and secondary consumers in an ecosystem.
5. Describe the sources of atmospheric water in the hydrologic cycle.
6. What are the major sources and sinks of atmospheric CO_2 in the carbon cycle?
7. What is nitrogen fixation, and why is it necessary for animals?
8. Why is phosphorus not found in the atmosphere?

Case Analysis

In 2017, an environmental group called Deep Green Resistance (DGR) filed suit in the United States District Court for the District of Colorado requesting that the court issue a declaratory judgment that the Colorado River Ecosystem "is a "person" capable of possessing rights; [and that] the Colorado River Ecosystem possesses the rights to exist, flourish, regenerate, be restored, and naturally evolve." The following is an excerpt from the plaintiffs' request for a declaratory judgment.

5. If we begin with water, we see—high in the sky—water dancing as vapor on wind currents. When the dance brings enough water together, clouds form. As clouds pass over the high Colorado Rockies, water freezes and falls as snow. Over the course of winter, clouds contribute their stores of water and snowpack builds. In spring, snowmelt forms creeks and streams who are guided by mountains through canyons and valleys. Rare summer rains do what they can to join the snowmelt.
6. Beneath the earth's surface, springs pull groundwater to form their own creeks and streams. Snowmelt, rain, and spring water intermingle with gravity. And, gravity gathers these waters as they tumble down stone faces, run across tree roots, and seep into sand and soil. The snowmelt, spring water, and gravity build in power as they mix. They soften mountainsides, carve through red rock, and brave the deserts that seek to exhaust them.

. . .

9. The Colorado River Drainage Basin is the seventh largest drainage basin in North America, covering 246,000 square miles. Ninety-seven percent (97 percent) of the drainage basin is in the United States. Twenty-five significant tributary rivers join the Colorado River, including the Green, Gila, San Juan, Little Colorado, Dolores, Gunnison, and Virgin Rivers.
10. Fourteen native fish lived in the Colorado River when European settlers arrived in the West, including four fish that are now endangered: the humpback chub, Colorado pikeminnow, razorback sucker, and bonytail. Only six known humpback chub populations persist. Colorado pikeminnow are no longer found below the Glen Canyon Dam. Wild populations of bonytail no longer exist. Endangered fish species with restricted ranges in Colorado tributaries include the Little Colorado spinedace, Kendall Warm Springs dace, desert pupfish, and springfish.
11. Springs that feed the Colorado, and the Colorado's tributaries, support several species of very rare snails, including the Overton assiminea, Grand Wash springsnail, Pahranagat pebblesnail, Moapa pebblesnail, and Hot Creek pebblesnail.
12. The Colorado River's natural communities include a diversity of forest and flora, including dense spruce-fir, pinyon-juniper, and mixed broadleaf and cottonwood forests; moist mountain grasslands where tufted hair grass, Thurber's fescue, and blue joint grass flourish; prolific willow carrs; desert scrublands; and sparse saltbush-greasewood basins.
13. The Colorado River's riparian communities are among the most important habitats for winged creatures in the Western United States. One hundred and thirty-nine (139) confirmed butterfly species can be found in Rocky Mountain National Park alone. Iconic, and endangered

or threatened, birds like the bald eagle, greater sage grouse, Gunnison sage grouse, peregrine falcon, yellow-billed cuckoo, summer tanager, and southwestern willow flycatcher make their homes in the Colorado River Watershed.

14. The scarcity of water in the deserts of the Southwest make the Colorado River Watershed vital for several amphibian species including the Colorado River toad, lowland leopard frog, and the relict leopard frog. Development and water diversion endanger these rare desert amphibians.
15. Many of the West's most recognizable mammals depend on the Colorado River Watershed for water and to sustain adequate food sources. Gray wolves, grizzly bears, black bears, mountain lions, coyotes, and lynxes walk the banks of the Colorado. Elk, mule deer, and bighorn sheep live in the Basin's forests. Beavers, river otters, and muskrats live directly in the River's flow as well as in streams and creeks throughout the Colorado River Basin.
16. The Colorado River provides water for close to 40 million people and irrigates nearly 4 million acres of American and Mexican cropland.

Questions

1. What evidence does DGR present that the Colorado River is an "ecosystem" as we understand the term?
2. We learned about the structure of the legal system in chapter 2. If the Colorado River Ecosystem were declared to be a "person" as DGR asks, would that mean that the ecosystem would have ***standing*** in the legal sense and could file lawsuits on its own behalf?
3. Are you convinced by DGR's argument? Why do you think that the Colorado River Ecosystem is or is not a "person" under U.S. law?
4. The DGR was ultimately unsuccessful, and the Circuit Court judge dismissed the DGR lawsuit in *Colorado River Ecosystem v. State of Colorado*.[25] This meant that the Colorado River Ecosystem is not a person under United States law. Was the judge correct in dismissing the lawsuit? Why or why not?
5. No federal court in the United States has given standing to an ecosystem to date, but several foreign countries have done so.[26] Should the United States follow these countries and declare that ecosystems have legal standing?

NOTES

1. For an excellent technical discussion of the question of whether viruses are alive or not, see: Koonin, E.V., & Starokadomskyy, P. 2016. Are Viruses Alive? The

Replicator Paradigm Sheds Decisive Light on an Old But Misguided Question. *Studies in History and Philosophy of Biological and Biomedical Sciences, 59,* 125–134. https://doi.org/10.1016/j.shpsc.2016.02.016. NIH National Library of Medicine: https://www.ncbi.nlm.nih.gov/pmc/articles/PMC5406846/.

2. A good discussion of these biomes is on Arizona State University's *Ask a Biologist* website: https://askabiologist.asu.edu/explore/biomes.

3. U.S. EPA: https://www.epa.gov/caddis-vol2/physical-habitat.

4. Polechová, J., & Storch, D. 2008. Ecological Niche. In: *Encyclopedia of Ecology,* 2. https://doi.org/10.1016/B978-008045405-4.00811-9.

5. See the articles in Wilson, E.O. (ed.). 1992. *Biodiversity.* The National Academies Press; Wilson, E.O. 1992. *The Diversity of Life.* W.W. Norton & Co.; and Snape, W. (ed.). 1996. *Biodiversity and the Law.* Island Press.

6. As we will see in our discussion of aquatic ecosystems, some deep-sea organisms use chemicals from deep in the Earth as a source of energy in a process called *chemosynthesis.*

7. The term *autotroph* translates to "self-eater," referring to the ability of plants to utilize photosynthesis to convert energy from the sun to convert CO_2 into glucose. This glucose is used as food by the plant but can also be used as a source of energy by *heterotrophs*, or "other eaters," which includes herbivores and carnivores.

8. The term *biomass* is used in ecology to mean the mass of living organisms in a given area or ecosystem at a given time and is expressed in units of weight (grams, kg, etc.). However, in other contexts, it can mean the dried (dead) weight of organisms in an area with all water removed.

9. This point is made quite elegantly by Dr. G. Tyler Miller in his famous quote: *Three hundred trout are needed to support one man for a year. The trout, in turn, must consume 90,000 frogs, that must consume 27 million grasshoppers that live off of 1,000 tons of grass* (American Chemist, 1971).

10. An excellent review of the significance of the water cycle to microbes is: Bertrand, J.C., et al. 2015. Biogeochemical Cycles. In: Bertrand, J.C., Caumette, P., Lebaron, P., Matheron, R., Normand, P., & Sime-Ngando, T. (eds.) *Environmental Microbiology: Fundamentals and Applications.* Dordrecht: Springer. https://doi.org/10.1007/978-94-017-9118-2_14.

11. See Ollivier, J., Töwe, S., Bannert, A., Hai, B., Kastl, E.-M., Meyer, A., Su, M.X., Kleineidam, K., & Schloter, M. 2011. Nitrogen Turnover in Soil and Global Change: Key Players of Soil Nitrogen Cycle. *FEMS Microbiology Ecology, 78*, 3–16.

12. Reid, W., Mooney, H., Cropper, A., Capistrano, D., Carpenter, S., & Chopra, K. 2005. *Millennium Ecosystem Assessment. Ecosystems and Human Well-Being: Synthesis.* Island Press.

13. 16 U.S.C. § 1533.

14. 16 U.S.C. § 1532(5)(A).

15. *Robertson v. Seattle Audubon Soc'y* (1590–1596), 503 U.S. 429 (1992).

16. 16 U.S.C. Ch. 31 §§ 1361 et seq.

17. Infrastructure Investment and Jobs Act (Pub. L. 117–58, 135 Stat.429), Section VIII.

18. https://www.epa.gov/eco-research.

19. https://oceanservice.noaa.gov/ecosystems/.

20. https://www.fws.gov/program/ecological-services.

21. https://www.fs.usda.gov/research/managingland/ecosystem.

22. https://www.nps.gov/subjects/parkscience/ecosystems.htm.

23. California Fish and Game Code, Ch. 10. §§ 2800–2835. Natural Community Conservation Planning Act.

24. https://floridadep.gov/eco-pro.

25. See: https://climatecasechart.com/case/colorado-river-ecosystem-v-state-colorado/.

26. For example, in 2018, the Administrative Court of Boyacá in the country of Colombia in South America recognized the Páramo in Pisba, a high Andean ecosystem facing significant mining, as a "subject of rights." The International Center for the Rights of Nature maintains a website, Advancing Legal Rights of Nature: Timeline, that looks at legal rights of nature at: https://celdf.org/advancing-community-rights/rights-of-nature/rights-nature-timeline/.

Chapter 4

Endangered Species, Biodiversity, and Ecosystems

Chapter Learning Outcomes

At the end of this chapter, the reader will be able to:

1. *Summarize the approach of early Europeans toward the environment and contrast it with the approach of indigenous people at the time.*
2. *Distinguish between the approaches of the conservationist and preservationist movements.*
3. *Discuss the requirements of Section 7 of the Endangered Species Act.*
4. *Describe the application of the takings provision of the Endangered Species Act.*
5. *List the steps in the Section 4 listing process of the Endangered Species Act.*
6. *Identify a federal law other than the ESA that protects ecosystems.*
7. *Describe the CITES convention.*

We explored some of the issues associated with ecosystems in chapter 3, but let's take a deeper dive into the issues associated with protecting the most vulnerable organisms in nature.

4.1 A BRIEF HISTORY OF SPECIES AND HABITAT PROTECTION

The relationships between humans and the natural world have not always been comfortable. It is undeniable, however, that wildlife and the natural areas they inhabit have had a profound impact on people since our earliest history as a species.[1] Evidence of this impact exists from the earliest Paleolithic petroglyphs and cave drawings to the most modern wildlife art. Professor Rodgers, in his popular environmental law textbook from the 1990s, said "[t]he human relationship to wildlife touches the deepest wellsprings of human emotion and behavior . . . [and] forges the best thinking about human ethics and morality."[2]

Indigenous peoples from around the world have evidently lived in harmony with nature and incorporated natural themes into their artwork and verbal traditions for millennia. However, when colonists from industrialized Europe came to North America, they brought with them the attitude that humans were intended to dominate nature and to use nature specifically to benefit themselves. These views of nature, often incorporated into what was known as the "pioneer spirit," dominated American attitudes from colonial times until the 1960s. A statement attributed to colonial clergyman and author Cotton Mather expresses early pioneer attitudes: "What is not useful [in nature] is vicious." Europeans, including American colonists, historically viewed wildlife as either a resource to be exploited or a danger to be removed (figure 4.1).

Mostly because of these early pioneer attitudes, early U.S. federal laws were directed only to preserve economically valuable hunting and fishing resources. Most regulation of wildlife and what we call "conservation" today was based on state laws. The first federal law to protect wildlife (however

Figure 4.1 A Nineteenth-Century Bear Hunt in California. Scene in the Frezno Grove of Mammoth Trees. *Source*: https://www.yosemite.ca.us/library/scenes_of_wonder_and_curiosity/mariposa_big_trees.html

indirectly) was the Lacey Act of 1900, which provides federal sanctions for interstate transportation of fish or wildlife taken in violation of federal, state, or foreign laws.[3] It has been frequently amended, but the Lacey Act remains a useful tool to prevent the overexploitation of wildlife resources.[4] In 1918, the Migratory Bird Treaty Act[5] "g[ave] the Secretary of the Interior power to adopt regulations for the protection of migratory birds [and] concentrated on establishing refuges for wildlife."[6] These laws were designed to protect wildlife as a resource to be exploited, however, not as a mechanism for protecting wildlife.

Attitudes about how Americans regarded wildlife and the relationship between humans and the land began to change in the nineteenth century. The emergence of the "environmental ethic" (Aldo Leopold, Garrett Hardin,

Figure 4.2 George Perkins Marsh in 1862. *Source*: Public domain image. https://commons.wikimedia.org/wiki/File:George_Perkins_Marsh.jpg

and others) led to increased public sensitivity to lost or degraded natural resources. As we discussed in chapter 1, during the 1870s and 1880s, George Perkins Marsh (figure 4.2) and Gifford Pinchot argued that advances in ecological theory could be used to allocate the earth's natural resources in a way that mutually benefited both humans and nature, which led to the development of the "conservationist" movement (i.e., careful management of the environment would ensure plentiful natural resources for the future).

Also during the 1870s and 1880s, John Muir (figure 4.3) and others advocated a form of natural morality that supported the preservation of large tracts of pristine wilderness. The "preservationist" movement resulted and was largely responsible for developing the nation's system of national parks and wilderness areas.

Figure 4.3 John Muir (1838–1914). *Source*: National Park Service. https://www.nps.gov/parkhistory/online_books/sontag/muir.htm

Attitudes toward wildlife had largely changed by the 1960s. During the decades after the end of World War II, the United States experienced heightened interest in environmental protection at all levels. Congress began to provide the kind of emphasis on the protection of species that exists today. The Endangered Species Preservation Act of 1966, an early version of the Endangered Species Act, authorized the Secretary of the Interior to identify "the names of the species of native fish and wildlife found to be threatened with extinction," and to acquire lands to protect threatened wildlife.[7] Unfortunately, the 1966 Act applied only to "native" fish and wildlife and was further limited in scope in that it applied only where "practicable."

4.2 THE ENDANGERED SPECIES ACT OF 1973

The federal ESA was passed in 1973. The position taken by Congress is expressed in the congressional finding that "various species of fish, wildlife, and plants in the United States have been rendered extinct as a consequence of economic growth and development untempered by adequate concern and conservation."[8] As we discussed in chapter 3, the ESA is probably the most important federal habitat and ecosystem protection law because an endangered or threatened species has a "critical habitat" that cannot be developed or converted to human use under most circumstances. It is one of the few environmental laws which place an absolute prohibition on destroying a resource, although the prohibition may not be absolute in all cases.

The ESA is divided into several sections. We will review the sections of the greatest importance to our discussion.

In Section 2 of the ESA, Congress made a strong statement that plant and animal species are important, and that they are worth protecting. ESA §2 includes statements that species of fish, wildlife, and plants have been so depleted in numbers that they are in danger of or threatened with extinction; [and] these species of fish, wildlife, and plants are of esthetic, ecological, educational, historical, recreational, and scientific value to the Nation and its people."[9]

ESA §2 also states the general purposes of the Act, which are "to provide a means whereby the ecosystems upon which endangered species and threatened species depend may be conserved, to provide a program for the conservation of such endangered species and threatened species."[10]

- ESA Section 7: The Jeopardy Provision and the Duty to Conserve

The Jeopardy Provision

Section 7 of the ESA states that every federal agency "shall . . . ensure that any action [by the agency] is not likely to jeopardize the continued existence of any endangered species or threatened species, or result in the destruction or adverse modification" of the critical habitat of the species.[11] It is important to note that Section 7 applies to all federal agency actions, but not to private individuals.

The duty imposed by §7 on federal agencies is reminiscent of the approach used by the NEPA that requires agencies to consider the environmental consequences of their actions, as we discussed in chapter 1.

The Duty to Conserve

Under Section 7 of the ESA, the Secretary of the Interior has an obligation to use programs which seek to conserve imperiled species "by carrying out programs for the conservation" of the imperiled species.[12] Of equal significance, however, is the mandate by ESA §4 that *all* federal agencies participate in conservation efforts:

"All other Federal agencies shall, in consultation with and with the assistance of the Secretary, utilize their authorities in furtherance of the purposes of this chapter by carrying out programs for the conservation of endangered species and threatened species listed pursuant to section 1533 [ESA §4] of this title."[13]

ESA §2(c)(1) reinforces this mandate by discussing congressional policy:

> *It is further declared to be the policy of Congress that all Federal departments and agencies shall seek to conserve endangered species and threatened species and shall utilize their authorities in furtherance of the purposes of this chapter.*[14]

Federal courts have interpreted the ESA §7 mandate strictly. One federal court famously held that the Department of the Interior must use "all methods which are necessary to bring any endangered . . . or threatened species to the point at which the [conservation measures under the ESA] . . . are no longer necessary."[15] In addition, the Department of the Interior "must do far more than merely avoid the elimination of protected species. It must bring those species back from the brink [of extinction] so that they may be removed from the protected class, and it must use all methods necessary to do so."[16]

ESA Section 9: The Taking Provision

Without question, ESA §9 is the most controversial provision in the ESA. Simply stated, ESA §9 states that "[I]t is unlawful for any person subject to the jurisdiction of the United States to . . . (B) take any [endangered or threatened] species within the United States or the territorial sea of the United States."[17]

The meaning of the word "take" is clarified by the ESA §3 definition: "The term 'take' means to harass, harm, pursue, hunt, shoot, wound, kill, trap, capture, or collect, or to attempt to engage in any such conduct." A controversy erupted in the early 1990s over what, exactly, was meant by the term "harm" in the definition of "take" in ESA §3. A series of property owners challenged the definition, arguing that the destruction of property that did not directly injure the animals themselves (red-cockaded woodpeckers and northern spotted owls) could not rise to the level of a "take" of a protected species in violation of ESA §9. After a series of inconsistent decisions by lower courts, the United States Supreme Court held that modification of critical habitat is "harm" under the statute.[18]

A similar problem has arisen with the definition for "harass" within the ESA §3 meaning of "take." As it appears in the USFWS regulations, "harass" is defined as:

Harass in the definition of "take" in the Act means an intentional or negligent act or omission which creates the likelihood of injury to wildlife by annoying it to such an extent as to significantly disrupt normal behavioral patterns which include, but are not limited to, breeding, feeding, or sheltering.[19]

Despite this regulatory definition, some federal courts have found that acts might not rise to the level of a "take" under ESA §9 because the ESA itself does not define "harass."[20]

ESA §9 goes even further. It is also unlawful for any person to "possess, sell, deliver, carry, transport, or ship, by any means whatsoever, any such species . . .".[21] This section is aimed primarily at halting the possession of animal "parts" when they are from an endangered or threatened species. Some defendants have argued that they are "innocent owners" of the parts in that they could not reasonably be expected to know that the parts are from a protected species. However, federal courts have often used a "strict liability" approach and held the defendants guilty of a "take" under the ESA regardless of their degree of fault.[22]

ESA §9 Applies to Private Individuals

Unlike ESA §7, which applies only to agencies, ESA §9's prohibition against the "taking" of protected species applies "to any person."[23] Courts have interpreted "person" loosely and have applied the ESA §9 "taking" prohibition to a variety of activities. For example, private hunters have been prohibited from selling protected bald eagle feathers,[24] and from shooting a grizzly bear[25] or a Florida panther,[26] even if the act was allegedly done to protect private property. On the other hand, it is not a "taking" if the person has a good faith belief that they are protecting themselves, a family member, or another individual from bodily harm by an endangered species.[27]

ESA Enforcement

Violations of ESA §9 are taken seriously. The most formidable enforcement is criminal. ESA §11(b)(1) states that "any person who knowingly violates any provision of this chapter, of any permit or certificate issued hereunder, or of any regulation . . ." is subject to criminal sanctions.[28] The key challenge for the government is proving that the defendant "knowingly" violated the ESA.[29] A person who is convicted of violating the ESA can "be fined not more than $50,000 or imprisoned for not more than one year, or both."[30]

However, the more common alternative to criminal enforcement are civil penalties. ESA §11(a) states that "any person who knowingly violates, and any person engaged in business as an importer or exporter of fish, wildlife, or plants who violates, any provision of this chapter, or any provision of any permit or certificate issued hereunder, or of any regulation . . . may be assessed a civil penalty by the Secretary of not more than $25,000 for each violation."[31] As with any civil trial, the burden of proof is lower in civil enforcement, and is therefore an easier burden on the agency.

In many cases, the agency's only remedy is the forfeiture by the defendant of the protected species or parts, or of the equipment used in collecting or transporting the protected species.

The ESA Section 4 Listing Process

Suppose that we have evidence that a species of animal is threatened by extinction in the future. What is the process for listing it as an endangered species under the ESA?

The current "listing" process is found in ESA §4.[32] The process begins when an interested person submits a petition to the Secretary of the Interior (or USFWS). Within ninety days, the Secretary must publish a finding of whether or not the petition contains sufficient scientific and commercial information to support action, and must then begin a status review.[33] Then, within 12 months, the Secretary must publish one of three findings (usually in the Federal Register): (1) the action is not warranted, (2) the action is warranted, or (3) the action is warranted, but is precluded by other pending listing proposals.[34] If the action is "warranted but precluded," the Secretary must demonstrate that "expeditious progress" is taking place to list qualified species.[35]

If the Secretary finds that the action is warranted, a notice must be published along with the proposed regulation, and a public hearing is held if one is requested. Within one year, the Secretary must adopt or withdraw the regulation.

Emergency listings can occur if the Secretary finds that a species is in "imminent peril."[36] These become effective as soon as published following a 240-day streamlined rulemaking procedure.

Endangered or Threatened?

At this point, the Secretary must determine if a listed species is "endangered" (in danger of extinction) or "threatened" (likely to become an endangered species in the foreseeable future). The Secretary must also determine the species' "critical habitat" to assure protection,[37] and implement a "recovery plan."[38] All may be challenged in federal court,[39] and will be discussed more fully below. There are over 1,300 species currently listed as either endangered or threatened.

Potential candidate species are placed into one of two categories while the listing process is completed. Species for which adequate data exist are listed as "category 1." About 600 U.S. species are in category 1. Species for which there are inadequate data are placed into "category 2," of which there are over 3,000 U.S. species. A species may be "delisted" (i.e., removed from the list) if: (1) it is extinct; (2) it has recovered to the point that it no longer requires ESA protection; or (3) it was originally listed based on erroneous data.[40] For example, in 1998, the bald eagle was reclassified from "endangered" to "threatened" (figure 4.4).

As we have seen, the ESA authorizes two levels of protection: "endangered" and "threatened." While the statute specifically authorizes the protection of species determined to be *endangered*, Congress delegated to the Secretary the authority to protect *threatened* species.[41] In fact, the Secretary has authorized virtually identical protection for *threatened* species in the regulations.[42]

What about species that have been proposed for listing as either endangered or threatened but for which the listing process is not yet complete? Under ESA §7(a)(4), government agencies must "confer with the Secretary" before taking an action that might jeopardize a candidate species.[43] However, this level of protection has not always worked to the advantage of the candidate species and may even invite the destruction of a species before it becomes listed by identifying its location.

Under ESA §4(b)(7), the Secretary is authorized to issue emergency regulations "to prevent a significant risk to the well being" of a candidate species. This provision is rarely used, however, given the press of numerous candidate species.[44] In addition, the USFWS regularly publishes a list of all candidate species and monitors their status. Presumably, any candidate species that is threatened would justify emergency regulations. Despite these safeguards, candidate species do *not* enjoy the same level of protection as listed species.

Figure 4.4 The American Bald Eagle. They were reduced from endangered to threatened status in 1998. The recovery of the bald eagle is remarkable, but relatively few endangered species experience a recovery like this. *Source*: Wikimedia Commons

Critical Habitat

Under ESA §4(a)(3),[45] the Secretary must designate critical habitat at the same time that a species is listed. "Critical habitat" is defined in ESA §3(5) as the "specific areas within the geographical area occupied by the species" that are essential to its conservation or that require special protection, or areas outside the area that the Secretary finds to be essential for its conservation.[46] As we discussed in chapter 3, the critical habitat of an endangered species is often used as a substitute for protecting an ecosystem.

There is an element of cost-benefit analysis in the designation of critical habitat. Section 4(b)(2) states that the Secretary:

"shall designate critical habitat . . . after taking into consideration the economic impact, the impact on national security, and any other relevant impact, of specifying any particular area as critical habitat. The Secretary may exclude any area from critical habitat if he determines that the benefits of such exclusion outweigh the benefits of specifying such area . . . unless he determines . . . that the failure to designate such area as critical habitat will result in the extinction of the species concerned."[47]

Many critical habitat designations have been delayed (often because of a lack of scientific data), and there have been many legal challenges.[48]

Recovery Plans, Plants, and Citizen Suits

Under ESA §4(f), the Secretary must also create a recovery plan for each listed species, unless a recovery plan will not promote the conservation of the species.[49] The ESA amendments of 1988 added that the recovery plan must be prepared "without regard to taxonomic classification," because it was found that recovery plans were only being prepared for the so-called "warm fuzzies" (like bald eagles, wolves, etc.). At this point, over 1,200 recovery plans are completed (or in process) for over 1,600 listed species.

Plants receive considerably less protection under the ESA than do "fish and wildlife." Originally, plants received no protection at all, but the 1982 amendments added a new §9(a)(2), which makes it unlawful to "remove or reduce to possession," or "maliciously damage or destroy," any protected plant species from land under federal jurisdiction.[50] However, the ESA's prohibition regarding plants does not extend to privately owned land.

ESA §11(g) contains a citizen suit provision that allows "any person" to commence a civil suit in federal District Court on his or her own behalf to force compliance with ESA mandates.[51] A plaintiff is limited to injunctive or declaratory relief under ESA §11(g), although damages may be pursued under common law. A successful plaintiff may be awarded attorney's fees.

4.3 DOES THE ENDANGERED SPECIES ACT WORK?

As we have seen, the impact of the ESA is significant. However, it's fair to ask whether the ESA's efforts have worked. Are vulnerable species in better shape with the ESA's mandates than they would have been without it? The answer depends on your perspective.

There is reason to be optimistic. Currently, there are more than 1,600 animals and plants that are listed as endangered or threatened in the United

States, although there are hundreds more waiting for protection. More specifically, it's estimated that in the absence of the ESA, at least 227 species might have gone extinct since the enactment of the legislation.[52]

Unfortunately, a vulnerable species may not experience the protection that was contemplated by the ESA. A 2022 Columbia Climate School study found that population sizes of species at the time of ESA listing did not significantly change between 1985–1991 and 1992–2020. The study also found long wait times before species receive protection, which further increases the risk of extinction to species with already small or rapidly declining populations. An additional problem is that funding allocations declined between 2010 and 2020, while the number of species listed for protection increased by over 300 percent during that time. As a result, protection has dropped by nearly 50 percent per species since 1985.[53]

The ESA has suffered from political troubles from the beginning. In recent years, the Trump Administration attempted to reduce the protection for species designated as threatened and redefined critical habitat to make it easier for private property owners to resist critical habitat designations. However, the Biden Administration quickly moved to restore these protections. It is likely that the ESA will continue to see politically motivated changes for the foreseeable future.

4.4 OTHER SPECIES, ECOSYSTEM, AND BIODIVERSITY LAWS

The federal ESAis important in protecting vulnerable species and organisms, but it is not the only federal law with this goal. The Marine Mammal Protection Act of 1972 (MMPA),[54] protects marine mammals (like dolphins and manatees) in a manner reminiscent of the Endangered Species Act. For example, a person may not "take" a marine mammal (where "take" is more broadly defined than in the ESA), even if the animal is not actually harmed.[55] It is also unlawful to possess or import marine mammals that were taken unlawfully (even if they were taken in a foreign country), if they were taken in an inhumane manner, or if they are taken when they are pregnant or nursing.[56]

The Bald Eagle and Golden Eagle Protection Act of 1962[57] is designed to protect our national symbol and its closest relative. Interestingly, golden eagles were added to the Bald Eagle Protection Act of 1940 only because juvenile bald eagles were being killed accidentally by hunters who thought they were golden eagles. Prohibited acts include "take" (defined broadly), as well as removing or destroying nests, eggs, or "parts" such as feathers. There are limited exemptions to protect livestock, falconry, and native American religious purposes, among others.

The Wild Free-Roaming Horses and Burros Act of 1971 protects those organisms in the U.S.[58] Ironically, horses and burros are not natural

Figure 4.5 Wild Horses Like These Are Protected under the Wild Free-Roaming Horses and Burros Act. *Source*: Public domain picture. Bureau of Land Management. https://www.blm.gov/blog/2020-09-10/top-6-things-know-about-wild-horse-range-pastures

inhabitants of the U.S. They were originally brought to what is now the United States by European colonists but have become an established part of many environments, especially in the western United States (figure 4.5).

In addition, most states have their own versions of the ESA, and most maintain a list of species of plants and animals for that state.[59]

4.5 INTERNATIONAL VULNERABLE SPECIES AND BIODIVERSITY LAWS

We discussed international environmental law in chapter 2. These agreements certainly involve international efforts to protect vulnerable species and biodiversity. Probably the most important such agreement is CITES, or the Convention on International Trade in Endangered Species of Wild Fauna and Flora (figure 4.6).

CITES began as a resolution adopted in 1963 at a meeting of members of the International Union for the Conservation of Nature (IUCN). Its aim is to ensure that international trade (import/export) in specimens of animals and plants included under CITES does not threaten the survival of the species in the wild (figure 4.6). This is achieved via a system of permits and certificates. CITES is a multilateral treaty to protect endangered plants and animals from the threats of international trade. As of 2023, there are 184 signatory countries, including the United States, which protect over 38,000 species.

Figure 4.6 CITES is an International Agreement that Protects Vulnerable Species around the World. *Source*: NOAA. https://cdhc.noaa.gov/outbreak-investigation/permitting/

Figure 4.7 Guns like This "Punt Gun," a Giant Shotgun, Are Prohibited under the Migratory Bird Treaty Act. *Source*: Public domain image. https://www.thevintagenews.com/2016/04/26/punt-gun-the-gigantic-vintage-shotgun-killed-90-waterfowls-at-once-2/

The Migratory Bird Treaty Act of 1918[60] protects birds that migrate into and out of the United States by implementing treaties between the United States and several other countries (Great Britain on behalf of Canada, Mexico, Japan, and the "Soviet Union"). The Act does not prohibit all hunting of migratory birds but rather limits the methods (e.g., trapping, snaring, poisoning, and the use of certain types of guns is prohibited (figure 4.7).

We will encounter other laws and international agreements that indirectly affect wildlife and ecosystems as we discuss other kinds of environmental protection.

* * *

REVIEW QUESTIONS

1. Describe the approaches to the environment of European settlers in North America.

2. Contrast the approaches of the preservationist and conservationist movements and describe their impact on U.S. environmental policy.
3. Describe and contrast the Jeopardy Provision and the Duty to Conserve in §7 of the ESA, and explain to whom they apply.
4. What does it mean to *take* an endangered species?
5. What are the defenses to a charge of taking an endangered species under §9 of the ESA?
6. What are the differences between *endangered* and *threatened* status under §4 of the ESA?
7. Explain what is meant by *critical habitat* under §4 of the ESA.
8. Does the Endangered Species Act work? Explain why it does or does not.
9. Describe two federal statutes other than the ESA that provide protection to vulnerable species, including what organisms they protect, and how the protection works.
10. Describe two different international agreements that protect vulnerable species.

Case Analysis

In 2018, the U.S. Supreme Court was asked to resolve an issue regarding the designation of critical habitat for an endangered species. The issue involved the designation of critical habitat for the dusky gopher frog (*Rana sevosa*) in Mississippi. The frog lives mainly underground in old-growth pine forests. Most of its original habitat has been destroyed by forestry processes. In declaring the frog an endangered species, the U.S. Fish and Wildlife Service designated critical habitat that included an area where the frogs still survive, but also included a parcel that was degraded, not currently inhabited by the frogs, and was not useful as habitat without major restoration. The Center for Biological Diversity, an environmental group, argued that the unoccupied habitat was still "critical habitat."

The Supreme Court was asked to determine if critical habitat for an endangered species had to be "habitat" where, presumably, the endangered species currently exists.

Chief Justice Roberts delivered the opinion of the Court. The following is an extract:

> According to the ordinary understanding of how adjectives work, "critical habitat" must also be "habitat." Adjectives modify nouns—they pick out a subset of a category that possesses a certain quality. It follows that "critical habitat" is the subset of "habitat" that is "critical" to the conservation of an endangered species.

Of course, "[s]tatutory language cannot be construed in a vacuum," and so we must also consider "critical habitat" in its statutory context. Section 4(a)(3)(A)(i), which the lower courts did not analyze, is the sole source of authority for critical-habitat designations. That provision states that when the Secretary lists a species as endangered he must also "designate any habitat of such species which is then considered to be critical habitat." 16 U. S. C. §1533(a)(3)(A)(i). Only the "habitat" of the endangered species is eligible for designation as critical habitat. Even if an area otherwise meets the statutory definition of unoccupied critical habitat because the Secretary finds the area essential for the conservation of the species, Section4(a)(3)(A)(i) does not authorize the Secretary to designate the area as critical habitat unless it is also habitat for the species.

The Center for Biological Diversity contends that the statutory definition of critical habitat is complete in itself and does not require any independent inquiry into the meaning of the term "habitat," which the statute leaves undefined. Brief for Intervenor-Respondents 43–49. But the statutory definition of "critical habitat" tells us what makes habitat "critical," not what makes it "habitat." Under the statutory definition, critical habitat comprises areas occupied by the species "on which are found those physical or biological features (I) essential to the conservation of the species and (II) which may require special management considerations or protection," as well as unoccupied areas that the Secretary determines to be "essential for the conservation of the species." 16 U. S. C. §1532(5)(A). That is no baseline definition of habitat—it identifies only certain areas that are indispensable to the conservation of the endangered species. The definition allows the Secretary to identify the subset of habitat that is critical, but leaves the larger category of habitat undefined.

Questions

1. What exactly did the Supreme Court decide? Can an unoccupied parcel be part of the critical habitat?
2. The Court relies heavily on its definition of "habitat." Is that the correct way to interpret the meaning of critical habitat under §4 of the ESA?
3. Suppose that the USFWS restored the area to its original condition before proposing it as critical habitat. Even though no frogs were present, the area was suitable for habitation. Would that have changed the Court's decision? Why, or why not?
4. Did the Supreme Court reach the correct decision under the circumstances of the case? Explain your answer.

NOTES

1. For an excellent discussion of the relationship between humans and nature by a famous ecologist, see Wilson, E.O. 1984. *Biophilia*. Harvard University Press.

2. Rodgers, W.H. 1994. *Environmental Law* (2nd ed.). West Publishing, at § 9.9.

3. 16 U.S.C. §§3371–78, and 18 U.S.C. §42.

4. See Baur, E.D., & Irvin, W.R. (eds.). 2010. *Endangered Species Act: Law, Policy, and Perspectives* (2nd ed.). American Bar Association. This book is a collection of essays by authorities on the Endangered Species Act that gives useful perspectives on the history and implementation of the ESA.

5. 16 U.S.C. 703 et seq. See Rholf, D. 1989. *The Endangered Species Act: A Guide to Its Protection and Implementation*. Stanford Environmental Law Society.

6. *TVA v. Hill*, 437 U.S. 153 at n. 20 (1978).

7. Pub. L. No. 89–669, 80 Stat. 926 (1966). See *TVA v. Hill*, 437 U.S. 153, 175 (1978).

8. 16 U.S.C. §1531(a)(1). See Coggins, S. 1993. An Ivory Tower Perspective on Endangered Species. *Natural Resources & Environment, 8*(1), 3.

9. 16 U.S.C. §1531(a).

10. 16 U.S.C. §1531(b). Note the reference to ecosystems.

11. 16 U.S.C. §1536(a)(2). See *Bennett v. Spear*, 117 S.Ct. 1154 (1997).

12. 16 U.S.C. §1536(a)(1).

13. See *Carson-Truckee Water Conservation District v. Clark*, 741 F.2d 257, 261 (9th Cir. 1984), cert. denied sub nom. *Nevada v. Hodel*, 470 U.S. 1083 (1985) [all federal agencies must carry out programs for the conservation of endangered and threatened species].

14. 16 U.S.C. §1531(c)(1).

15. *National Wildlife Federation v. Hodel*, 23 ER Cases 1089, 1092 (E.D. Cal. 1985) [prohibiting the Secretary of the Interior from allowing hunters to use lead shot due to poisoning of bald eagles].

16. *Defenders of Wildlife v. Andrus*, 428 F. Supp. 167, 170 (D.D.C. 1977) [disallowing pre-dawn game bird hunting since poor visibility might result in endangered species being shot by accident].

17. 16 U.S.C. §1538(a)(1).

18. *Babbitt v. Sweet Home Chapter of Communities for a Greater Oregon*, 115 S.Ct. 2407 (1995).

19. 50 CFR §17.3.

20. see *U.S. v. Hayashi*, 5 F.3d 1278 (9th Cir. 1993) [tuna fishermen were not guilty of a "take" of endangered porpoises under the MMPA by firing rifles in the water, because the Act does not define the term "harass"].

21. 16 U.S.C. §1538(a)(1)(D).

22. See *U.S. v. One Handbag of Crocodilus Species*, 856 F.Supp. 128 (E.D.N.Y., 1994) [An "innocent owner" of protected crocodile hide items is still liable under the ESA, since a "strict liability" approach is necessary to effect congressional intent].

23. 16 U.S.C. §1538(a)(1).

24. *U.S. v. Dion*, 762 F.2d 674 (8th Cir. 1985), *rev'd in part on other grounds*, 476 U.S. 734 (1986).

25. *U.S. v. St. Onge*, 676 F. Supp. 1041 (D. Mont. 1987).

26. *U.S. v. Billie*, 667 F. Supp. 1485 (S.D. Fla. 1987).

27. 16 U.S.C. §1540(b)(3).

28. 16 U.S.C. §1540(b)(1).

29. See *U.S. v. Billie*, 667 F.Supp. 1485 (S.D. Fla. 1987); and *U.S. v. Nguyen*, 916 F.2d 1016 (5th Cir. 1990).

30. 16 U.S.C §1540(b)(1). Updated sentencing laws have increased the maximum fine to not more than $100,000 and increased jail sentences. See Sobeck, E. 1993. Enforcement of the Endangered Species Act. *Natural Resources & Environment, 8*(1), 30.

31. 16 U.S.C. §1540(a)(1). Again, updated sentencing laws have increased the maximum civil penalties.

32. 16 U.S.C. §1533.

33. 16 U.S.C. §1533(b)(3)(A), see *Marbled Murrelet v. Babbitt*, No. 91-522R (W.D. Wash, 1993) [the USFWS was ordered to issue a rule proposing critical habitat using the best scientific information then available, although the USFWS argued insufficient information].

34. 16 U.S.C. §1533(b)(3)(B).

35. 16 U.S.C. §1533(b)(3)(B)(iii)(II).

36. 16 U.S.C. §1533(b)(7).

37. 16 U.S.C. §1333(b).

38. 16 U.S.C. §1533(f).

39. See *Trinity County Concerned Citizens v. Babbitt*, No. 92–1194 (D.D.C. 1993) [unsuccessful challenge to Northern spotted owl critical habitat designation].

40. 50 CFR §424.11(d).

41. 16 U.S.C. §1533(d); see *Christy v. Hodel*, 857 F.2d 1324 (9th Cir. 1988), *cert. den'd sub nom. Christy v. Lujan*, 109 S.Ct. 3176 (1989).

42. 50 CFR §§17.31(a) and 17.71(a).

43. 16 U.S.C. §1536(a)(4); see *Wilson v. Block*, 708 F.2d 735 (D.C. Cir. 1983), *cert. den'd.*, 104 S.Ct. 371 (1983).

44. 16 U.S.C. §1533(b)(7).

45. 16 U.S.C. §1533(a)(3).

46. 16 U.S.C. §1532(5). See also the Department of the Interior web page at: https://www.doi.gov/ocl/critical-habitat.

47. 16 U.S.C. §1533(b)(2). The U.S. Supreme Court examined this section of the ESA in *Weyerhaeuser Co. v. U.S. Fish and Wildlife Service*, 586 U.S. ___; 139 S.Ct. 361 (2018).

48. For example, see *Northern Spotted Owl v. Lujan*, 758 F.Supp. 621 (E.D. Wash. 1991).

49. 16 U.S.C. §1533(f); see *Nat'l Wildlife Federation v. Nat'l Park Svc.*, 669 F. Supp. 384 (D. Wyo. 1987); and *Sierra Club v. Yeutter*, 926 F.2d 429 (5th Cir. 1991).

50. 16 U.S.C. §1538(a)(2).

51. 16 U.S.C. §1540(g).

52. Greenwald, N., et al. 2019. Extinction and the U.S. Endangered Species Act. *PeerJ, 7*, e6803. https://doi.org/10.7717/peerj.6803.

53. Eberhard, E.K., Wilcove, D.S., & Dobson, A.P. 2022. Too Few, Too Late: U.S. Endangered Species Act Undermined by Inaction and Inadequate Funding. *PLOS ONE, 17*(10), e0275322. https://doi.org/10.1371/journal.pone.0275322.

54. 16 U.S.C §§1361–1407.

55. See *Katelnikoff v. Dept. of the Interior,* 657 F.Supp. 659 (D. Alaska 1986); and Fed'n of Japan Salmon Fisheries Coop. *Ass'n v. Baldridge*, 679 F.Supp. 37 (D.D.C. 1987), *aff'd sub nom. Kokechik Fisherman's Ass'n v. Sec. of Commerce*, 839 F.2d 795 (D.C. Cir. 1988).

56. See *Animal Welfare Inst. v. Kreps*, 561 F.2d 1002 (D.C. Cir. 1977); and *Globe Fur and Dyeing Corp. v. U.S.*, 467 F.Supp. 177 (D.D.C. 1978), *aff'd*, 612 F.2d 586 (D.C. Cir. 1988).

57. 16 U.S.C. §668.

58. Wild Free-Roaming Horses and Burros Act, 16 USC Ch. 30, as amended.

59. Camacho, A.E., Robinson-Dorn, M.J., Yildiz, A., & Teegarden, T. 2017. Assessing State Laws and Resources for Endangered Species Protection. *Environmental Law Reporter, 47*(10, UC Irvine School of Law Research Paper No. 2017–51). https://ssrn.com/abstract=3060882.

60. 16 U.S.C. §§703–711.

Chapter 5

Public Land and the Environment

Chapter Learning Outcomes

At the end of this chapter, the reader will be able to:

1. *Distinguish between public and private land.*
2. *Describe the responsibilities of federal agencies in managing public lands.*
3. *Discuss the concepts of the Tragedy of the Commons effect, externalities, and maximum sustainable yield in managing public land.*
4. *Identify customary uses by private parties on public lands.*
5. *Describe the management of public land outside the United States.*
6. *Describe and contrast clear-cutting, selective cutting, and ecologically sustainable forestry.*
7. *Discuss the role of fire management in forestry management.*

As we saw in chapter 4, the distinction between private and public land is important when we consider the reach of federal regulation of land use. In this chapter, we will explore what we mean by public land, explore where public lands are located, and finally, we will take a close look at an example of how environmental resources are managed on public forests.

5.1 WHAT IS PUBLIC VERSUS PRIVATE LAND?

Americans are famously protective of their land. This is partly a result of our founding as a nation when we rebelled in part against a government that was less than supportive of private property rights. But what, exactly, do we mean by public versus private land?

First, public land in the United States is land that is *held in trust* for the American people by the federal government and managed by federal agencies. In our legal system, holding property in a trust relationship means that a trustee (federal agencies in this case) holds a position of trust in relation to the beneficiaries (citizens), and the trustee must act in the beneficiaries' best interest. In other words, agencies of the U.S. government have a responsibility to the citizens of the United States to protect and manage public land in a way that benefits all citizens. Unfortunately, how we perceive our best interests varies from person to person, and there are many disputes over whether public land is being managed properly or not.

As trustees, federal agencies have responsibilities to the public far beyond environmental stewardship. We will explore these agencies and their responsibilities in the next sections and look more extensively at land use practices.

Private land, on the other hand, is simply land that is not owned by the government. The owner of private land may be a U.S. citizen, a corporation or other entity, or even a resident of a foreign country under some circumstances. But it is even more complicated than that because states and municipalities own property that is usually considered to be public but has a special relationship both with the federal government and private citizens.

For land that is privately owned, property rights may be more restrictive than people think. Americans often believe that private property rights are absolute when they are not. Let me use myself as an example. I own an acre of land in unincorporated (meaning not part of a town or city) Charleston County, in South Carolina. So, surely, I am an owner of private property. But does that mean that I have complete control over my property so that I can do what I want? It does not. For one thing, I can't burn my yard waste, fire a gun, or set off explosives on my property without permission, because these actions are controlled by local laws that protect my neighbors from my irresponsible actions. Furthermore, I am required to grant an easement across my property to the local power company, meaning that they have limited access to my property so they can inspect and repair their equipment.

In many cases, a local government may have an easement on the portion of your property that abuts your street, an alley, or a sidewalk. Property owners who own land adjacent to a public beach or other public amenity may be required to permit access to the public across their property. And, if you live in a community with a homeowners' association, then you may have restrictions on the appearance of your house, the vegetation in your yard, or even the color of your mailbox. All of this is to say that Americans rarely enjoy the absolute ownership of private land that we think we do. So, it should not be surprising that certain environmental restrictions might apply to private property, such as the critical habitat requirement under the Endangered Species Act that we discussed in chapter 4.

We will see other examples of incursions on private property rights in subsequent chapters. But for now, let's focus on public lands in the United States.

5.2 PUBLIC LAND IN THE UNITED STATES

We said above that public land is managed by federal agencies. But what agencies are involved? The most significant of these federal agencies are as follows:

- Bureau of Land Management (BLM)
- U.S. National Park Service (NPS)
- Bureau of Reclamation
- U.S Fish and Wildlife Service (USFWS)
- U.S. Forest Service
- National Oceanic and Atmospheric Administration (NOAA)
- U.S. Department of Defense (U.S. Army Corps of Engineers)

These agencies have a huge responsibility. Approximately 42 percent of the land in the United States is publicly owned, with 25 percent of the nation's land owned by the federal government.

Figure 5.1 shows the distribution of federal public land in the United States. Note the high concentration of public land in western states, which reflects the fact that western states were added to the United States later in our history and were less subject to private ownership. In fact, some states, like Nevada, Utah, and Alaska, are over 80 percent federal land. These states are well known for disputes involving private uses of public land.

We can understand our federal public lands more thoroughly by understanding their history. Most federal lands were acquired early in U.S. history, from purchases (like the Louisiana Purchase from France), or wars (like the Spanish-American War), or treaties made with foreign countries. In the beginning, the federal government used this land to encourage growth, settlement, and economic development.

Land that was not developed, homesteaded, or sold in the early years remained in federal ownership as public land. Land in the eastern United States was converted to private ownership much earlier in the process, so less eastern land remained in the federal system compared to the western United States.

Today, principles of land use planning and environmental protection preserve the natural resources and scenic beauty found on public land. Let's explore some of these principles.

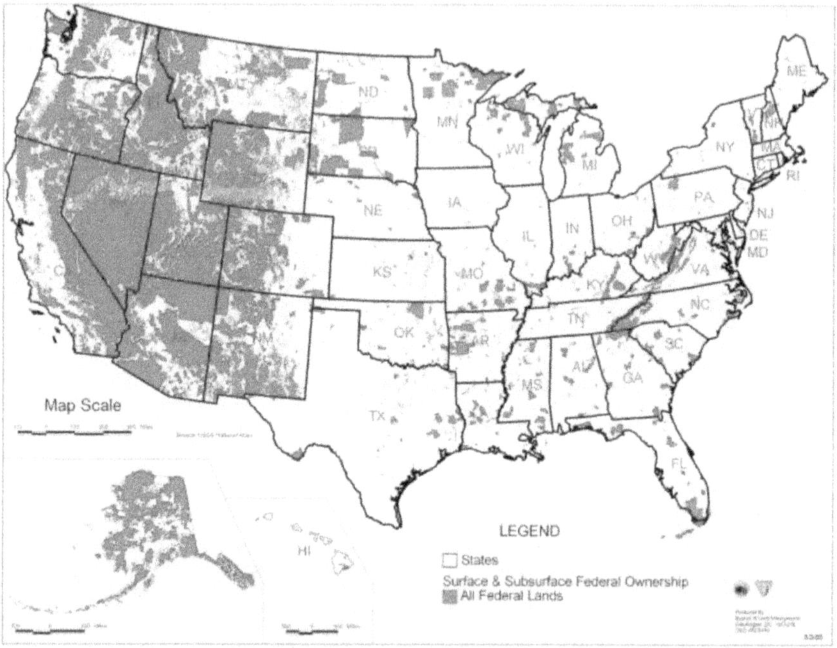

Figure 5.1 Federal Public Land in the United States. Note the heavy concentration of public lands in western states. *Source*: Bureau of Land Management—http://wilderness.org/blm-lands

Land Use on Public Land

We have determined that federal agencies are the stewards of public land. This stewardship must be based on the rules of land use planning, understanding that every human use of the land alters it in some way.

Let's explore some of the principles of land use planning as the concept is used by federal agencies. As we do this, please understand that people do not always agree on land use planning and management priorities. We will examine some of the points of disagreement as we go.

Understanding land use and management issues requires the use of three key concepts:

1. The Tragedy of the Commons effect
2. Externalities
3. Maximum sustainable yield

Let's examine these three concepts.

Tragedy of the Commons Effect

We discussed the Tragedy of the Commons[1] effect in chapter 1. Recall that the issue was the breakdown of altruism, based on the observation that "cheaters" will inevitably take advantage of voluntary restrictions on the use of the Global Commons for their own benefit. It is important to keep this principle in mind as we examine policies for the management and planning of public land use by federal agencies.[2]

Externalities

The concept of externalities was originally developed by economists, but it has broad applications to environmental biology and law. In general, an *externality* is a side effect or consequence of an industrial or commercial activity that affects other parties without this being reflected in the cost of the goods or services involved. A classic example of an externality is the pollination of surrounding crops by bees kept for honey. This is a *positive externality* because it conveys a benefit (free pollination) that is not reflected in the price of honey.

Environmental biology and law are concerned with negative externalities because of the environmental damage for which no one bears the cost. An example of a *negative externality* would be the air pollution produced by a coal-fired electricity generation plant, which has a negative effect on health and the environment, but the health and environmental cost is not passed on to consumers.

Environmental policy has considered the effects of negative externalities for many years. Effective land-use management and planning should also consider externalities for the proper management of public land.

Maximum Sustainable Yield

The concept of maximum sustainable yield has its origin in the field of economics, but it has become an important part of the science of ecology, as well as agriculture, forestry, and fisheries and wildlife management. It has become an integral part of the process of land use management and planning.

In population ecology and economics, *maximum (or optimum) sustainable yield (MSY)* is the level of effort (LOE) that maximizes the difference between total revenue and total cost in a system. Of course, revenue and cost have very different meanings in different fields of study.

A useful, simple example of MSY comes from fisheries management. In fisheries, MSY is defined as the maximum catch (in numbers or mass) that can be removed from a population over a specific (or indefinite) period of time without destroying the population. The concept of MSY relies on the surplus production generated by a population that is depleted below its environmental carrying capacity.

A convenient way to understand MSY is to examine the growth curves used by many scientists, from economists to ecologists. A growth curve is a graph that plots the growth in population size of a population of living things over time. The graph plots *population size (N)* over *time (t)* (figure 5.2). The graph shows us that a population will usually grow at a slow rate early in time, but the rate will rapidly increase to a maximum. Think of a small group of ants that are introduced into an environment that has no ants. If there are ample food and shelter resources and little competition, the population will grow rapidly. This rapid growth period is often called *exponential growth*, and it is seen in many populations of organisms, such as disease organisms that infest a new host, or weeds in a field that has been recently plowed or disturbed.

There is, however, a limit on the size a population can reach (i.e., the number of individuals in a population). There will always be some limiting factor in the environment that serves to prevent a population from reaching an infinite size. This upper limit is known as the *carrying capacity (K)*, and all populations have one. As population growth approaches the carrying capacity over time, the rate of increase in growth begins to decrease as population size approaches K. This produces the characteristic sigmoid (S-shaped) curve that we see in figure 5.2. This sigmoid growth curve represents what we call *logistic growth* in a population.[3]

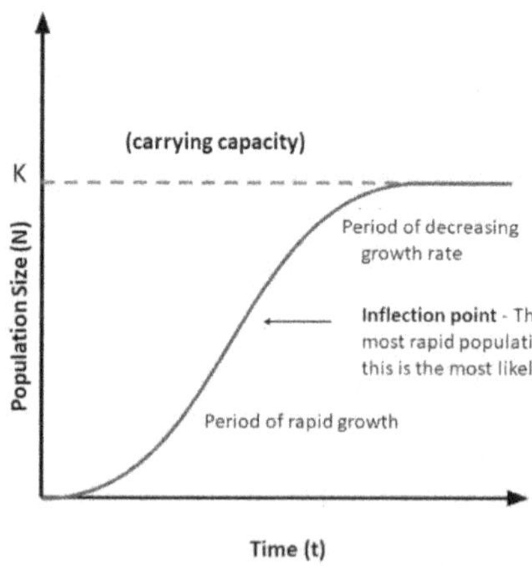

Figure 5.2 The Logistic Growth Curve. Population growth is rapid during the early time periods and reaches its greatest rate of population increase at the inflection point, which is also the optimal population size for MSY. As the curve approaches the carrying capacity, growth rate decreases as it approaches K. Modified from Nchisick (own work).
Source: Wikimedia Commons. ttps://creativecommons.org/licenses/by-sa/4.0/deed.en

The information in a logistic growth curve can be used to determine the Maximum Sustainable Yield for an organism. As seen in figure 5.2, the time of most rapid growth in population size is the best estimate of MSY. The idea is that we can remove individuals from a population either for commercial purposes (e.g., the fishing industry) or for environmental purposes (e.g., experimental or recovery purposes), and the population can still recover.

As we will see, MSY is an important concept in everything from commercial exploitation of a species to critical habitat design, to recovery plans for vulnerable species.

A relevant question is where the human population of the world would fall on the graph? There is some disagreement, but we are probably near the top of the graph where it closely approaches the carrying capacity (K). We have not reached K yet because humans have a unique ability to raise the carrying capacity, K. Modern agriculture, medicine, and a variety of technological innovations have increased the carrying capacity and permitted more humans to live on the planet under generally better conditions than in the past.

5.3 WORLDWIDE LAND USE

Before we begin investigating land use management in the United States, let's investigate how land is used worldwide. Figure 5.3 gives us a broad view of worldwide land use, but it does not tell the whole story. From the perspective of human beings, it is useful to consider how much of the world's land is used for agriculture. According to figure 5.3, 7 percent of the world's land is cropland and 27 percent is livestock (including livestock feeds), for a total of 34 percent devoted to agriculture (crops and livestock).

Given this information, it is useful to consider how agricultural land use has changed over the centuries. Figure 5.4 shows changes in the total land used for agriculture in different parts of the world from the year 1600 until 2016. Worldwide, total agricultural land has increased from less than one billion hectares in 1600 to over five billion hectares in 2016. This is an astonishing increase in agricultural land use, but it begs the question of what was lost to accommodate the increase. The uncomfortable answer is that valuable environmental resources (wildlife, ecosystems, etc.) have been lost so that humans will have more food, and the trend continues. We will examine this and related issues elsewhere in this book.

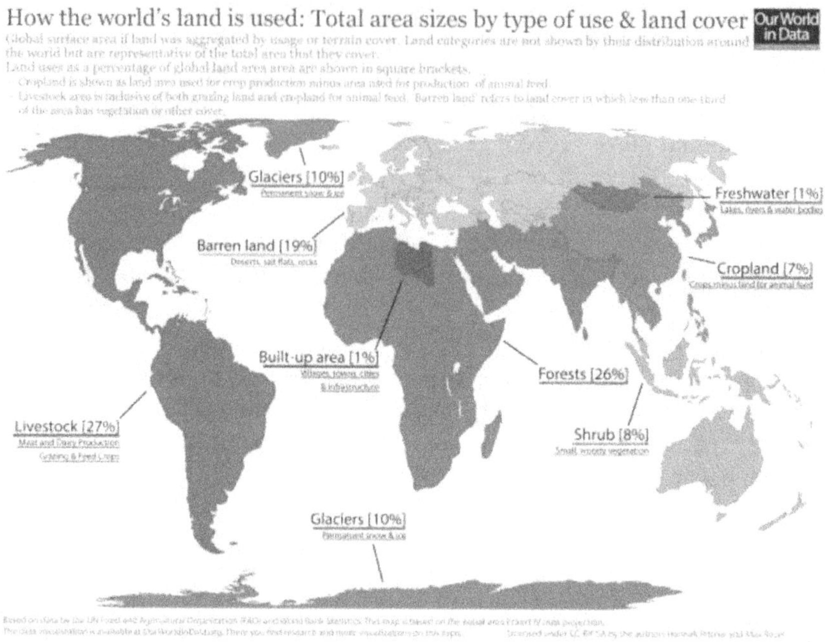

Figure 5.3 Land Use Worldwide. *Source*: Ritchie, H. and Roser, M. (2013)—"Land Use". Published online at OurWorldInData.org. Retrieved from: 'https://ourworldindata.org/land-use' [Online Resource]

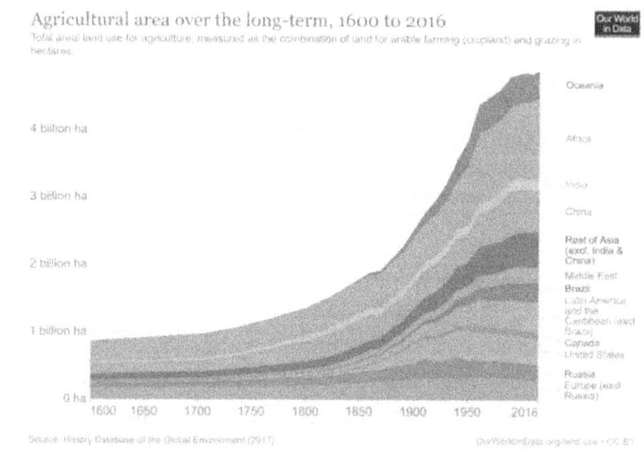

Figure 5.4 Worldwide Agricultural Land Use, 1600–2016. This graph shows changes in total agricultural land use for the regions of the planet from the year 1600 until 2016. Note that the Americas do not become significant agricultural areas until the 1800s. *Source*: Klein Goldewijk, K., A. Beusen, J.Doelman and E. Stehfest (2017), Anthropogenic land use estimates for the Holocene; HYDE 3.2, Earth System Science Data, 9, 927–953. https://ourworldindata.org/land-use

A good deal of the land in figure 5.4 is not agricultural, and this is the land that we hope is managed for environmental purposes. The United Nations recognizes six categories of public land uses:

- **National parks** are managed for scientific, educational, and recreational use, and sometimes for their beauty or unique landforms.
- **Managed Resource Protected Areas** are designated for the sustained use of biological, mineral, and recreational resources.
- **Habitat/Species Management Areas** are actively managed to maintain biological communities.
- **Strict Nature Reserves and Wilderness Areas** are set aside to protect species and ecosystems.
- **Protected Landscapes and Seascapes** permit non-destructive use of natural resources while allowing for tourism and recreation.
- **National Monuments** are designated to protect unique sites of special natural or cultural interest.

Note that many land uses under the U.N. classification permit use by people for recreation, tourism, and even the extraction of resources! As we will see, very few national or international resources exclude humans.

5.4 LAND MANAGEMENT ON PUBLIC LAND

As you might expect, very different kinds of land management are used for different environmental conditions. Table 5.1 shows data on land uses in the contiguous forty-eight United States. Different federal agencies have developed technical management skills that provide them with the expertise to

Table 5.1 U.S. Land Use in 2017

Use	*Acreage (M)*	*km2 (M)*	*Total (%)*
Pasture/range	654	2.647	35
Forest	538.6	2.18	28
Cropland	391.5	1.584	21
Special use*	168.8	0.683	9
Miscellaneous*	68.9	0.279	4
Urban	69.4	0.281	4
Total**	1,891	7.653	100

(M) is millions

* *Special use* areas include national parks (29 M acres) and state parks (15 M), wildlife areas (64.4 M), highways (21 M), railroads (3 M), military bases (25 M), airports (3 M), and a few others. *Miscellaneous* includes cemeteries, golf courses, marshes, deserts, and other areas of "low economic value".

** Total for forty-eight contiguous states.

Data after Merrill, D. and Leatherby, L. (July 31, 2018). "Here's How America Uses Its Land". Bloomberg. https://www.bloomberg.com/graphics/2018-us-land-use/

manage parts of public lands, for example, the United States Forest Service has primary responsibility for the management of the nation's forests, the BLM oversees much of the open land in western states, and the U.S. Fish and Wildlife Service oversees land where fish and wildlife are managed.

Uses of public lands is a highly controversial topic that has taken on political overtones. These uses are often historic, but in many cases they adapt to current conditions.

Historically, public land uses were regulated by the following:

- **Resource conservation ethic:** The belief that people should maximize the use of resources, based on the greatest good for everyone.
- **Multiple-use lands:** A U.S. classification used to designate lands that may be used for recreation, grazing, timber harvesting, and mineral extraction.

In fact, public lands in the United States are usually available for many private uses, including:

- *Recreation*—This may be more or less restrictive, but usually includes wildlife watching and hiking, to fishing and hunting. Vehicle uses may be limited.
- *Grazing of livestock*—Usually restricted to leases, often controversial.
- *Timber harvesting*—Heavily regulated, requires leases.
- *Oil and gas drilling*—Requires leases. It is very controversial.
- *Mining*—Also requires leases. A common source of litigation.

The only public land that prohibits most private uses are designated Wilderness Areas, which were created by the 1964 Wilderness Act.[4] The Wilderness Act protects over 9.1 million acres (37,000 km^2) of federal public land.

Section 2 of the Wilderness Act provides a justification for and definition of what constitutes an area of land as wilderness. Wilderness Act land is chosen from existing federal land and by determining which areas are considered to meet the following criteria:

- Minimal human imprint
- Opportunities for unconfined recreation
- At least five thousand acres
- Educational, scientific, scenic, or historical value
- Have no commercial enterprises within them, or any motorized travel, or other form of mechanical transport (e.g., vehicles, motorcycles, bicycles).

Wilderness areas allow some private human activities, but these activities are far more restricted than under other statutes.

Wilderness areas in the United States are shown in figure 5.5.

Figure 5.5 Wilderness Areas in the United States. *Source*: U.S. Department of Agriculture, Forest Service. https://data.fs.usda.gov/geodata/other_fs/wilderness/index.php

As we discussed earlier, different federal agencies are responsible for the management of federal public land across the United States. However, more than 95 percent of all federal lands are managed by just four federal agencies:

- Bureau of Land Management (BLM): grazing, mining, timber harvesting, and recreation
- U.S. Forest Service (USFS): timber harvesting, grazing, and recreation
- National Park Service (NPS): recreation and conservation.
- U.S. Fish and Wildlife Service (USFWS): conservation, hunting, and recreation.

It is these agencies that have the greatest management expertise. How they manage their resources depends on the nature of the environment. Discussion of all land use management by all agencies is well beyond the scope of this book, but let's look at the example of national forests for illustrative purposes.

Land Use Management of Forests

Management issues differ for rangelands, forests, and parks. We will examine forest management practices as a specific example of public land use management (figure 5.6).

Figure 5.6 Francis Marion National Forest, in South Carolina. The forest is managed for a variety of uses, including logging, hunting, fishing, and hiking. Note the fire-scarred tree trunks that result from controlled burns. *Source*: Daniel Barcelona, own work. Wikimedia Commons. https://creativecommons.org/licenses/by-sa/3.0/deed.en

Forest management is the primary responsibility of the U.S. Forest Service, an agency of the U.S. Department of Agriculture. The Forestry Service describes its modern mission this way: "The overriding objective of the Forest Service's forest management pro-gram is to ensure that the National Forests are managed in an ecologically sustainable manner."[5]

But it was not always this way. The national forests were originally intended as working forests with multiple objectives, specifically to improve and protect the forests, to secure favorable watershed conditions, and to furnish a continuous supply of timber for the use of citizens of the United States. Timber harvest was then (and still is) done primarily by private companies for profit.

Forest management objectives have since expanded and evolved to include ecological restoration and protection, research and product development, fire hazard reduction, and the maintenance of healthy forests. Forest Service managers today use timber sales, as well as other vegetation management techniques such as prescribed fire, to achieve modern Forest Service objectives.[6] These activities have captured substantial public attention and, in some cases, become hotly debated issues.

Figure 5.7 Clear-Cutting of a Forest near Eugene, Oregon. *Source*: Calibas, own work. Wikimedia Commons. https://creativecommons.org/licenses/by-sa/3.0/deed.en

Silviculture

Management of forests for timber production is called *silviculture*, and it takes several forms:

- **Clear-cutting** removes most, if not all, trees from an area and is often coupled with replanting. The resulting trees are then all the same age. This is a particularly destructive practice and can seriously alter ecosystem values for centuries (figure 5.7).

Clear-cutting not only destroys nearly all ecosystem values at the time, but these forests are often replanted with trees of the same age and species in what is known as a tree farm. These tree farms provide little opportunity for healthy forest diversity of plants and animals.

Clear-cutting is a less common practice today than it was in the past, but it is still used by National Forest managers to restart forest successional stages on seriously disturbed land.

- In **selective cutting**, also known as *selection cutting,* single trees or small numbers of trees are harvested. The resulting forest consists of trees of varying ages (figure 5.8). This practice retains more of the forest ecosystem values than clear-cutting by providing trees of varying ages and sizes as a source of food, shelter, and nesting sites, but it still results in a forest ecosystem that is quite different from the natural ecosystem that would exist in the absence of all logging.

Figure 5.8 Selective Cutting of Ponderosa Pine in an Oregon Forest. *Source*: U.S. Forest Service archive. https://www.fs.usda.gov/rds/imagedb/results?place%5B1%5D=Arizona&place%5B2%5D=Fort+Valley+Experimental+Forest&collection%5B0%5D=Fort+Valley+Experimental+Forest+historical+photographs&year%5B0%5D=1939

Figure 5.9 An East Texas Sustainable Forest. Note the diversity of tree sizes and other plant diversity. *Source*: USDA Sustainable Forestry and Land Retention Workshop. https://www.nrcs.usda.gov/events/sustainable-forestry-and-land-retention-workshop

- **Ecologically sustainable forestry** is an approach to removing trees from forests in ways that do not unduly affect the viability of other trees (figure 5.9). This approach is increasingly popular with Forest Service management, as well as private forest managers all over the world. Sustainable forestry provides much healthier forests with far greater biodiversity of plants and animals than the alternatives.

Fire Management

The history of fire management in American forests is long, varied, and often uncomfortable. Our history tells us that indigenous people in the United States have used controlled burning as part of their environmental management for centuries (figure 5.10). But these techniques would not be employed by the U.S. government until the 1960s.

Fire management in national forests prior to the 1960s was based almost entirely on preserving the logging industry as a critical part of national interests. Forest rangers and the public alike were trained to spot and extinguish fires as quickly and completely as possible. In fact, a 1935 U.S. Forest Service's fire management policy stipulated that all wildfires were to be suppressed by 10 a.m. the morning after they were first spotted under any circumstances.[7] Generations of Americans grew up with Smokey the Bear's entreaties that "only YOU can prevent forest fires."

Figure 5.10 *The Grass Fire*. This famous 1908 painting by American artist Frederic Remington showing nineteenth-century Native Americans using controlled burns to manage prairie resources. *Source*: Public Domain image. Wikimedia Commons. https://commons.wikimedia.org/wiki/File:Remington_The_Grass_Fire_1908.jpg

This all began to change in the 1960s and 1970s when ecologists began to study the important role of fire in maintaining the natural integrity of many ecosystems worldwide, especially prairie and forest ecosystems. Ecologists explained that many important plants (and animals) were adapted to periodic low-intensity burning. These periodic fires would burn away dead grasses and understory plants, while allowing the adapted grasses and trees to thrive. In fact, it was determined that many forest and grassland ecosystems required periodic, controlled burning to maintain their ecosystem values.

Beginning in the 1960s, federal agencies began utilizing the controlled burns that are used to manage almost all national forests today. These managed burns are a common sight in many parts of the country and are recognized as a critical part of maintaining the health of forest ecosystems to preserve natural values as well as timber harvests.[8]

Prescribed burns, like that in figure 5.11, is a fire deliberately set under controlled conditions to reduce the accumulation of dead biomass on a forest floor. The trunks of the trees are resistant to fire damage, and the flames never reach the more vulnerable crowns of the trees. Prescribed burns like these help to reduce the risk of uncontrolled natural fires.

Sadly, many forests in the United States suffer from decades of fire suppression without modern fire management, which has left them vulnerable to wildfires. A series of droughts, made worse by climate change, has left many forests vulnerable, especially in western states (figure 5.12). Forest fires in the western United States and Canada have become so common over the past several decades that dealing with the loss of ecosystems, property, and lives

Figure 5.11 A Prescribed Burn in a New Jersey Forest. U.S. Forest Service. *Source*: USDA Photo by Phil Graeve. https://www.fs.usda.gov/features/professionals-prepare-for-prescribed-burn

Figure 5.12 The Elk Bath Fire. A wildfire on the East Fork of the Bitterroot River on the Sula Complex in the Bitterroot National Forest in Montana, U.S., in 2000. *Source*: Public domain image. John McColgan, Bureau of Land Management, Alaska Fire Service. Wikimedia Commons.https://en.wikipedia.org/wiki/Wikipedia:File_copyright_tags

has become almost an annual event. A 2024 study found that the worldwide frequency of wildfires doubled in just twenty years.[9]

But there is certainly hope for the future. As forests are destroyed by fire, they are usually replaced by well-managed forest ecosystems.

As you can see, the management of public lands is complicated and requires advanced expertise by the federal agencies that are involved.

* * *

Review Questions

1. What is "public land" as the term is used in the United States?
2. Describe the roles of the Forest Service, the Bureau of Land Management, and the National Park Service in managing the nation's public lands.
3. Distinguish between a positive and a negative externality and explain their relationships to environmental protection.
4. What is maximum sustainable yield, and how does it factor into management of public land and resources?
5. Distinguish between clear-cutting, selective cutting, and ecologically sustainable forestry as silviculture techniques, and explain their impacts on public land and resources.
6. Describe the role of prescribed burns in forestry management.

Case Analysis

In 2024, the U.S. Supreme Court decided the case *Sheetz v. El Dorado County*. George Sheetz, a property owner in El Dorado, California, wanted to build a small, prefabricated home on his residential parcel of land. However, to obtain a permit, he was required to pay a $23,420 fee to mitigate local traffic congestion, a standard fee in the county not specifically for the impact of his property alone. Mr. Sheetz challenged the fee as an unlawful "exaction" of money under the Takings Clause of the Fifth Amendment to the U.S. Constitution, arguing that the fee amount should be only what was necessary to offset traffic congestion attributable to his specific property development. The County's predetermined fee schedule, Sheetz argued, failed to meet that requirement. The following is an extract from Justice Barrett's decision for the Supreme Court.

> When the government wants to take private property to build roads, courthouses, or other public projects, it must compensate the owner at fair market value. The just compensation requirement comes from the Fifth Amendment's Takings Clause, which provides: "nor shall private property be taken for public use, without just compensation." By requiring the government to pay for what it takes, the Takings Clause saves individual property owners from bearing "public burdens which, in all fairness and justice, should be borne by the public as a whole." *Armstrong* v. *United States*, 364 U.S. 40, 49 (1960).
>
> Our decisions in Nollan and Dolan address this potential abuse of the permitting process. There, we set out a two-part test modeled on the unconstitutional conditions doctrine. First, permit conditions must have an "essential nexus" to the government's land-use interest. The nexus requirement ensures that the government is acting to further its stated purpose, not leveraging its permitting monopoly to exact private property without paying for it. Second, permit conditions must have "'rough proportionality'" to the development's impact on the land-use interest. A permit condition that requires a landowner to give up more than is necessary to mitigate harms resulting from new development has the same potential for abuse as a condition that is unrelated to that purpose. This test applies regardless of whether the condition requires the landowner to relinquish property or requires her to pay a "monetary exactio[n]" instead of relinquishing the property.
>
> . . .
>
> The California Court of Appeal declined to assess the County's traffic impact fee for an essential nexus and rough proportionality based on its view that the Nollan/Dolan test does not apply to "legislatively prescribed monetary fees." That was error. Nothing in constitutional text, history, or precedent supports exempting legislatures from ordinary takings rules.
>
> The Constitution's text does not limit the Takings Clause to a particular branch of government. The Clause itself, which speaks in the passive voice,

"focuses on (and prohibits) a certain 'act': the taking of private property without just compensation." It does not single out legislative acts for special treatment. Nor does the Fourteenth Amendment, which incorporates the Takings Clause against the States. On the contrary, the Amendment constrains the power of each "State" as an undivided whole. §1. Thus, there is "no textual justification for saying that the existence or the scope of a State's power to expropriate private property without just compensation varies according to the branch of government effecting the expropriation." Just as the Takings Clause "protects 'private property' without any distinction between different types," it constrains the government without any distinction between legislation and other official acts. So far as the Constitution's text is concerned, permit conditions imposed by the legislature and other branches stand on equal footing.

The same goes for history. In fact, special deference for legislative takings would have made little sense historically, because legislation was the conventional way that governments exercised their eminent domain power. Before the founding, colonial governments passed statutes to secure land for courthouses, prisons, and other public buildings. These statutes "invariably required the award of compensation to the owners when land was taken." Colonial practice thus echoed English law, which vested Parliament alone with the eminent domain power and required that property owners receive "full indemnification . . . for a reasonable price." [quotation sources omitted]

. . .

In sum, there is no basis for affording property rights less protection in the hands of legislators than administrators. The Takings Clause applies equally to both—which means that it prohibits legislatures and agencies alike from imposing unconstitutional conditions on land-use permits.[10]

Questions

1. What exactly did the Supreme Court decide? Does Mr. Sheetz still have to pay a fee? How much?
2. How did the Court balance the Takings Clause of the Fifth Amendment against the Police Power of the state of California? Does the Fifth Amendment limit the state's power to police its people?
3. Did the state's financial charge to Mr. Sheetz amount to a "taking" under the Fifth Amendment? After all, the state did not physically remove Mr. Sheetz's property from his possession. How did the Supreme Court respond to this issue?
4. The Supreme Court discussed the *Nollan/Dolan* two-part test. The names come from two previous Supreme Court decisions. What is the two-part Nollan/Dolan test for Fifth Amendment takings?

NOTES

1. Hardin, G. 1968. The Tragedy of the Commons. *Science, 162,* 1243–1248. https://doi.org/10.1126/science.162.3859.1243.

2. As a reminder, the concept of the Tragedy of the Commons is not universally accepted.

3. Logistic growth has been shown in many populations of organisms. Any good ecology textbook will have numerous examples of logistic growth of organisms in nature (for example, see: Molles, M., & Sher, A. 2022. *Ecology: Concepts and Applications*. McGraw-Hill).

4. 16 U.S.C. ch. 23 § 1131 et seq.

5. U.S. Forest Service web site at: https://www.fs.usda.gov/forestmanagement/aboutus/index.shtml.

6. For more information, visit the National Forest Service web site at: https://www.fs.usda.gov/forestmanagement/aboutus/index.shtml.

7. For an excellent review of early fire management, see: U.S. Forest Service. 2001. *Evolution of Federal Wildland Fire Management Policy. Review and Update of the 1995 Federal Wildland Fire Management Policy*. National Park Service, U.S. Forest Service, January. https://web.archive.org/web/20070810191055/http://www.nifc.gov/fire_policy/docs/chp1.pdf.

8. An excellent video has been produced by North Carolina State University on fire ecology and forest succession. https://www.youtube.com/watch?v=W_ntvTEben4.

9. Cunningham, C.X., Williamson, G.J., & Bowman, D.M.J.S. 2024. Increasing Frequency and Intensity of the Most Extreme Wildfires on Earth. *Nature Ecology and Evolution*. https://doi.org/10.1038/s41559-024-02452–2.

10. *Sheetz v. El Dorado County*, 601 U.S. ___ (2024).

Chapter 6

Air, Atmosphere, and Climate

Chapter Learning Outcomes

At the end of this chapter, the reader will be able to:

1. *Describe the four layers of the atmosphere and their characteristics.*
2. *List the important atmospheric gases and recognize which are variable gases and which are meteorologically significant.*
3. *Describe the effects of uneven heating, rotation, and other factors that are important in determining earth's weather.*
4. *Identify important surface issues on the earth that influence local weather patterns.*
5. *Contrast the origins and environmental effects of sulfurous and photochemical smog.*
6. *Discuss the six criteria air pollutants and their environmental and health effects.*
7. *Describe the origins and applications of National Ambient Air Quality Standards (NAAQS).*
8. *Contrast primary and secondary air quality standards.*
9. *Explain the origins and environmental effects of acid rain.*
10. *Describe the Clean Air Act's regulation of mobile sources of air pollution.*

We begin our investigation of the "media" of the environment (air, water, and land) by investigating the science and law of the *atmosphere*, the envelope of gases (air) surrounding the earth. Our approach is to describe the science of the atmosphere and climate, then investigate the laws that regulate and preserve them.

6.1 INTRODUCTION TO THE ATMOSPHERE

A good starting point is the basic structure of the earth's atmosphere. The atmosphere consists of five layers. Each of the layers is bounded by "pauses" where the greatest changes in thermal characteristics, chemical composition, movement, and density occur (figure 6.1).

Basic information on the five layers is in table 6.1.

The lowest layer of the atmosphere, the *troposphere*, is of particular significance to most ecosystems because it is the layer where most weather and climate occur. The troposphere is composed of the atmospheric gases that we call "air." The relative abundance of these gases is an important driver of ecosystem health, as we will discuss below.

Table 6.2 illustrates the abundance of tropospheric gases at sea level. We can see that the atmosphere is composed of a mixture of gases. These gases make up what is known as ambient air (that is, outdoor air). More specifically, "ambient air" is defined by the regulations under the CAA as "that portion of the atmosphere, external to buildings, to which the general public has access."[1] Some of these gases (indicated by "*" in table 6.2) are variable, depending on local or other atmospheric conditions.

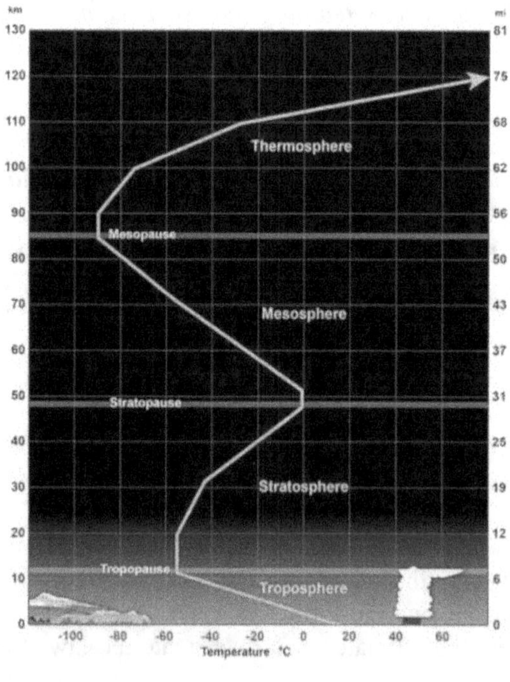

Figure 6.1 Lower Layers of the Atmosphere. Average temperature profile for the lower layers of the atmosphere. *Source*: National Oceanic and Atmospheric Administration. https://www.noaa.gov/jetstream/atmosphere/layers-of-atmosphere

Table 6.1 Characteristics of the Layers of the Atmosphere*

Each of the atmospheric layers is bounded by a "pause" where the greatest changes in thermal characteristics, chemical composition, movement, and density occur.

Exosphere

- Outermost layer of the atmosphere.
- Extends from about 375 miles (600 km) to 6,200 miles (10,000 km) above the earth's surface.
- Here, atoms and molecules escape into space; satellites orbit the earth.
- At the bottom of the exosphere is a transition layer called the *thermopause*.

Thermosphere

- Between about 53 miles (85 km) and 375 miles (600 km) above the earth.
- Known as the upper atmosphere.
- Gases of the thermosphere become increasingly dense closer to the earth.
- Incoming high-energy ultraviolet and x-ray radiation from the sun begins to be absorbed by the molecules in this layer, causing a large temperature increase.
- Because of this absorption, the temperature increases with height, from -184°F (-120°C) at the bottom of the layer to 3,600°F (2,000°C) near the top.
- The bottom of the thermosphere is the *mesopause*.

Mesosphere

- Between around 31 miles (50 km) above the earth to 53 miles (85 km).
- Gases continue to become denser as one descends.
- Temperatures increase as one descends, rising to about 5°F (-15°C) near the bottom of this layer.
- Both the stratosphere (the next layer down) and the mesosphere are considered the middle atmosphere.
- The transition boundary between the mesosphere and the stratosphere is the *stratopause*.

Stratosphere

- Between 4–12 miles (6–20 km) above the earth's surface to around 31 miles (50 km).
- Contains 19 percent of atmospheric gases but very little water vapor. Temperature increases with height. Heat is produced in the process of the formation of ozone, and this heat is responsible for temperature increases, from an average -60°F (-51°C) at the tropopause to a maximum of about 5°F (-15°C) at the top of the stratosphere. Warmer air is located above cooler air. This prevents convection as there is no upward vertical movement of the gases. As such, the location of the bottom of this layer is readily seen by the anvil-shaped tops of cumulonimbus clouds
- The transition layer at the bottom of the stratosphere is called the *tropopause*.

Troposphere

- Known as the lower atmosphere, almost all weather occurs in this region.
- Begins at the earth's surface, but the height varies. It is 11–12 miles (18–20 km) high at the equator, 5½ miles (9 km) at 50°N and 50°S, and just under four miles (6 km) high at the poles.
- As the density of the gases in this layer decreases with height, the air becomes thinner. Therefore, the temperature in the troposphere also decreases with height. The temperature decreases with increasing altitude from an average about 62°F (17°C) to -60°F (-51°C) at the tropopause.

Information after the National Oceanic and Atmospheric Administration. https://www.noaa.gov/jetstream/atmosphere/layers-of-atmosphere

Table 6.2 The Composition of Ambient Air at Sea Level

Gas Name	*Chemical Formula*	*Percent Volume at Sea Level % (15°C)*
Nitrogen	N_2	78.08
Oxygen	O_2	20.95
*Water	H_2O	0 to 4
Argon	Ar	0.93
*Carbon Dioxide	CO_2	0.0360
Neon	Ne	0.0018
Helium	He	0.0005
*Methane	CH_4	0.00017
Hydrogen	H_2	0.00005
*Nitrous Oxide	N_2O	0.00003
*Ozone	O_3	0.000004

* = Variable gases
Data after tornado.sfsu.edu/geosciences (7/13/18)

Although there are numerous gases that make up the ambient air of the atmosphere, the top four gases make up 99.998 percent of the volume of clean air. It is worth noting that five of these gases (nitrogen, oxygen, water, carbon dioxide, and ozone) contribute most significantly to the atmosphere and atmospheric conditions of greatest scientific importance.

By far the most abundant atmospheric gas is nitrogen (78 percent). Nitrogen dilutes oxygen and other gases and prevents rapid burning at the earth's surface, since oxygen gas is a necessary reactant of the combustion process. Nitrogen is also needed and used by organisms in the environment to make proteins, although as nitrogen gas, N_2, it is mostly unavailable to living things.

Oxygen is the second most abundant gas in ambient air. As we discussed in chapter 3, oxygen is a necessary part of aerobic respiration and metabolism for most living things. It is used to make complex molecules that are essential for life and is a necessary part of the process of oxidation of materials used in metabolism and growth.

The remaining gases are far less abundant. Water (H_2O) exists as vapor in most ambient air, although the relative volume varies greatly depending on atmospheric conditions. Atmospheric water constitutes between zero and 4 percent of atmospheric gas volume and is responsible for much of what we recognize as weather. Another atmospheric gas is argon, which is a relatively non-reactive gas used in light bulbs, in double-pane windows, and to preserve artifacts.

Carbon dioxide (CO_2) is an atmospheric gas that is essential for plants and other organisms to make sugars through photosynthesis. Photosynthesis is essential for other life as well because during photosynthesis, water molecules are split, and oxygen is released back into the atmosphere. Carbon dioxide is also among the greenhouse gases that trap radiant heat and prevent the escape of heat into outer space in what we recognize as the "greenhouse effect," explained below.

6.2 AN INTRODUCTION TO WEATHER AND CLIMATE

Our interactions with atmospheric gases are mostly through weather and climate. The primary difference between weather and climate is time. *Weather* is the condition of the atmosphere over a short period of time, usually measured in hours or days. *Climate* refers to atmospheric changes over longer periods of time, usually 30 years or more. A useful (but complicated) classification of worldwide climates is the Köppen-Geiger climate classification, shown in figure 6.2.

We all know that weather and climate vary significantly in different parts of the world. This is due to a variety of factors, but primarily to patterns of air circulation, ocean currents, and local topography that distribute heat and precipitation unevenly (figure 6.3). While there are many factors that contribute to the variety of climates, we can identify several significant factors that determine circulation in the lower atmosphere.

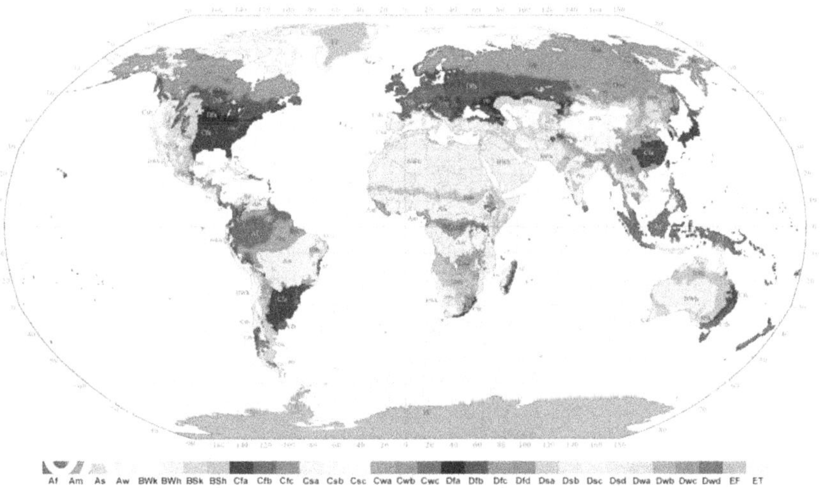

Figure 6.2 The Köppen-Geiger Climate Classification. Climate patterns based on precipitation and temperature patterns. *Source*: NOAA: https://www.noaa.gov/jetstream/global/climate-zones/jetstream-max-addition-k-ppen-geiger-climate-subdivisions

Rotation of the Earth on Its Axis

As the earth rotates, the equator travels faster than the polar regions. As a result, masses of heated air above the equator move north or south to cooler areas, but are deflected to the west or east over the planet's surface (figure 6.3). These patterns create the *prevailing winds*, or surface winds that blow almost continuously and help distribute air, moisture, and heat through the atmosphere. The prevailing winds are often known as the "trade winds" and have been used for ocean navigation by mariners for centuries. These and other factors cause the atmosphere to form large regions called *cells*, distinguished by the direction of air movement. Adjacent cells have opposite rotational wind patterns. These cells form conditions that are partly responsible for the location of many of the world's deserts just north or south of the equator at thirty degrees north or south latitude.

Uneven Heating of the Surface of the Earth by the Sun

Radiation from the sun strikes more directly at areas near the equator than areas closer to the poles, so air and surface temperatures tend to be much higher. This is why tropical areas near the equator tend to be much warmer than polar regions, and temperate regions have intermediate temperatures.

Many regions of the earth experience seasonal changes in climate, especially in temperature. The earth's spin axis is tilted with respect to its orbital plane, so that when the earth's axis points toward the sun, it is summer for

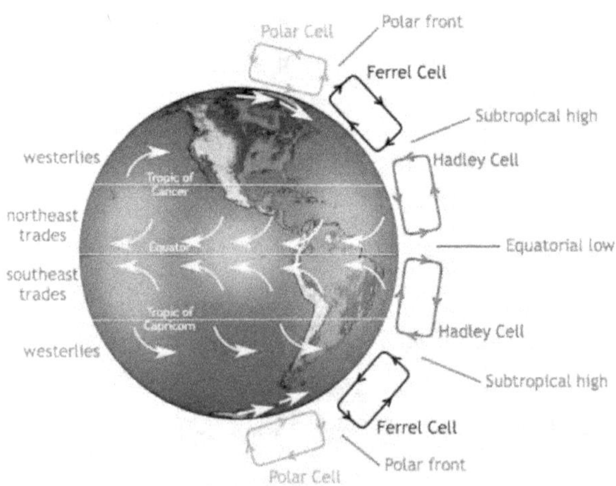

Figure 6.3 Patterns of Global Circulation. *Source*: NASA Earth Observatory. https://earthobservatory.nasa.gov/blogs/earthmatters/2017/

that hemisphere. When the earth's axis points away, it is winter. The farther an area is from the equator, the more indirectly the sun's rays strike during the winter, which creates extremely low temperatures near the poles.

Physical Properties of Air, Water, and Land

Radiant energy from the sun evaporates water from the surface of the earth, especially the oceans, and transfers heat from the surface to the atmosphere, especially near the equator. The evaporation of water creates large cyclical atmospheric convection cells that circulate water vapor, air, and heat both vertically and laterally in the atmosphere.

The earth's major ocean currents are driven by prevailing winds, the earth's rotation, heat, and the presence of landforms like continents. Tropical waters absorb more heat from the sun and the atmosphere, and differences in water density create both warm surface waters and deeper, colder waters. These mass movements of water redistribute heat across the earth's surface, and strongly influence climate, especially near coasts. Many of these currents, like the famous Humboldt Current in the eastern Pacific Ocean, are an important source of nutrient-rich, highly oxygenated ocean water which supply these resources to coastal ecosystems.

An important example of the interactions between the ocean and the atmosphere is the El Niño Southern Oscillation, or ENSO, phenomenon shown in figure 6.4. ENSO is a global weather phenomenon that is associated with a band of warm ocean water that develops in the central and east-central equatorial Pacific, including the area off the Pacific coast of South America. This shift can produce worldwide climate changes, causing flooding in some areas and drought in others, and can affect populations of marine species by changing water temperature, as well as the distribution and availability of nutrients.

The Greenhouse Effect

The greenhouse effect is familiar to anybody who has left a car in an open parking lot on a sunny day with the windows rolled up. After a few hours, the air in the car can reach dangerously high levels, many degrees higher than the outside ambient air. This results from ultraviolet (UV) radiation from the sun entering the car through the clear windows and heating the air and surfaces in the car. However, the heat is trapped in the car because the closed windows do not allow it to escape. The same principle works for glass-walled agricultural greenhouses that allow plants to grow inside even during the coldest months of winter.

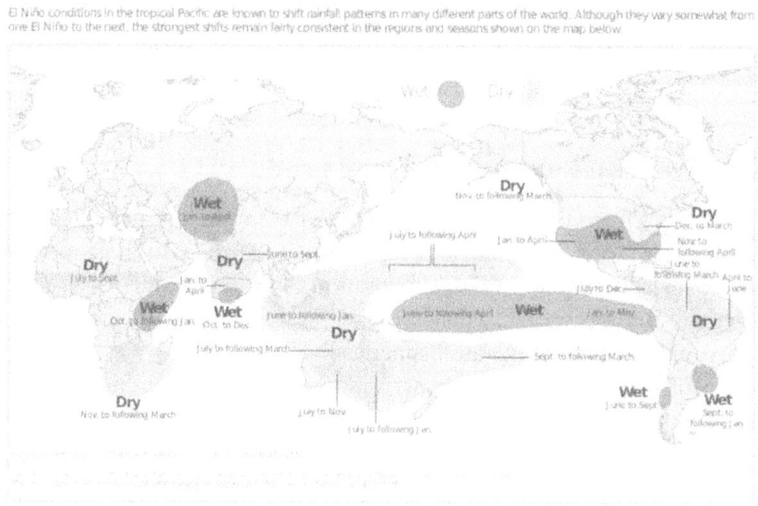

Figure 6.4 El Niño—Rainfall Events. Typical rainfall patterns during El Niño events. Such teleconnections are likely during El Niño events, but not certain. A strong El Niño will shift the Pacific jet stream, which helps determine temperature and precipitation patterns over the United States and alters the track of winter storms—leading to more winter storms over California and the southern United States. Record rainfall often strikes Peru, Chile, and Ecuador during an El Niño year. In contrast, El Niño can also cause severe droughts over Australia, Indonesia, and parts of southern Asia. *Source*: NOAA. https://www.noaa.gov/understanding-el-nino

Table 6.3 The Major Greenhouse Gases

Most of these are naturally occurring gases in the earth's atmosphere, but human activities can increase them dramatically, leading to excess atmospheric warming.

- Carbon dioxide (CO_2)
- Methane (CH_4)
- Water vapor (H_2O)
- Ozone (O_3)
- Nitrous oxide (N_2O)
- Industrial gases:
  - Hydrofluorocarbons (HFCs)
  - Perfluorocarbons (PFCs)
  - Sulfur hexafluoride (SF_6)
  - Nitrogen trifluoride (NF_3)

See U.S. Energy Information Administration. Energy and the environment explained. Greenhouse gases. https://www.eia.gov/energyexplained/energy-and-the-environment/greenhouse-gases.php

We mentioned above that carbon dioxide is a potent greenhouse gas, but there are others. Table 6.3 lists the major greenhouse gases.

In the atmosphere, greenhouse gases trap heat from the sun in much the same way as the windows of a car or the glass walls of a greenhouse (figure 6.5).

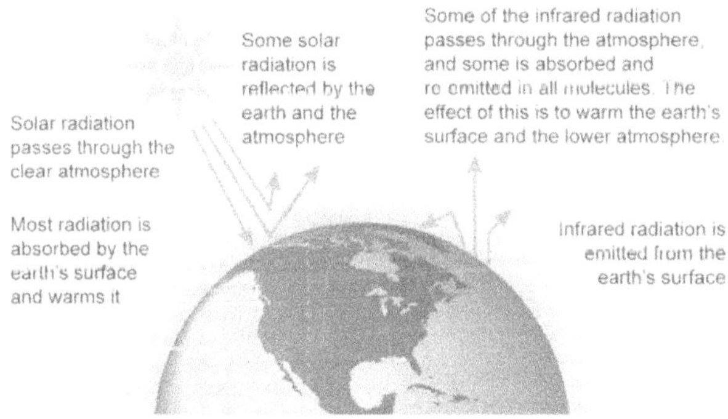

Figure 6.5 The Greenhouse Effect. Greenhouse gases trap heat in the lower atmosphere, warming the earth's surface and climate. *Source*: U.S. Energy Information Administration. Energy and the environment explained. Greenhouse gases. https://www.eia.gov/energyexplained/energy-and-the-environment/greenhouse-gases.php

The greenhouse effect is important in maintaining warmth at the surface of the earth, but it can also increase global temperatures to dangerous levels when greenhouse gases increase above normal levels. We will have more to say about greenhouse gases in chapter 12, our chapter on global climate change.

Topography and Landforms

Weather and climate patterns over land are influenced to a great degree by the elevation of the area. At higher elevations, atmospheric pressures are lower, temperatures are cooler, and water condensation may be greater, leading to fog and cloud formation. Lower elevations tend to be warmer and wetter, with important exceptions such as in deserts.

The presence of mountains often plays a key role in determining climate due to the *rain shadow* effect. Rain shadow deserts are created when mountain ranges lie parallel to moist, coastal areas. Prevailing winds moving inland force warm, moist air to rise over the mountains. As the air rises, pressure decreases, temperature drops, and water condenses from the moist air as precipitation. When the winds force air over the mountain crests and down the far side, they are very dry. Descending air makes it difficult for additional clouds and precipitation to form (figure 6.6). Without another source of moisture, rain shadow deserts are formed on the far side of these mountains. Deserts such as the Mojave and Chihuahuan Deserts in the United States, the Atacama Desert in South America, and the Tibetan Plateau in Asia are rain shadow deserts.

Figure 6.6 The Rain Shadow Effect. *Source*: National Park Service. https://www.nps.gov/jotr/learn/nature/deserts.htm

6.3 SMOG

Smog is a specific type of air pollution that depends on both climatic and topographic characteristics of an area. It is a combination of harmful pollutants (often appearing relatively low to the ground as a yellow-brown haze) that are introduced into the atmosphere by both natural and human-induced processes. It was first described in the 1940s in European cities like London as a mixture of smoke and fog (hence the name "smog"). However, the modern definition is far more complex.

Smog is made up of many chemicals, including nitrogen oxides (NO_x), sulfur dioxide (SO_2), carbon monoxide (CO), and volatile organic compounds (VOCs), but the two main components of smog are particulate matter (PM) and ground-level ozone (O_3).

Sulfur (or Sulfurous) smog (often called London-type smog) is primarily a product of burning large quantities of high-sulfur coal. Sulfur smog was responsible for the severe smog experienced by Londoners in the early twentieth century and is still common in many parts of the world.

Photochemical smog is a more modern phenomenon. It is produced when sunlight reacts with nitrogen oxides and at least one volatile organic compound (VOC) in the atmosphere. Nitrogen oxides come from sources like car exhaust, coal power plants, and factory emissions. VOCs are released from petroleum products like gasoline, paints, and many cleaning solvents. When sunlight contacts these chemicals, they form airborne particles and ground-level ozone (O_3), which is known as photochemical smog.

Landforms like mountains may create "bowls" in which smog is trapped, creating particularly dangerous conditions, especially when local weather conditions create inversions that further trap smog. Cities like Los Angeles and Beijing are well-known for their periodic smog conditions (figure 6.7).

Figure 6.7 Smog in Los Angeles, California. Smog is a particular problem where an area is surrounded by mountains, which trap air pollutants near the ground. *Source*: U.S. Environmental Protection Agency. https://www.epa.gov/sciencematters/how-science-protecting-air-we-breathe

Smog has serious effects on health and the environment. Ground-level ozone (a key component of photochemical smog) is associated with numerous health problems, including diminished lung function, increased hospital admissions and emergency room visits due to asthma, and increases in premature deaths.

Ground-level ozone can lead to reductions in agricultural crops and commercial forest yields, reduced growth and survivability of tree seedlings, and increased plant susceptibility to disease, pests, and other environmental stresses (such as harsh weather). Animal species that depend on plants for food resources are also affected.

6.4 THE ECOLOGY OF AIR AND ATMOSPHERE

We discussed ecosystems and their relationship to the environment in chapter 3. In particular, the major biogeochemical cycles (water, carbon, nitrogen, sulfur, and phosphorus cycles) provide major inputs into the global atmosphere. The relationship between ecosystems and the atmosphere is of critical importance to ecosystem health.

A substantial body of scientific research has demonstrated that ecosystems are impacted by air pollution, particularly sulfur and nitrogen emissions, and ground-level ozone as it affects their ability to function and grow. Emissions of both sulfur dioxide and nitrogen oxides react with water in the atmosphere to form sulfur and nitrogen acids, which fall as "acid rain." Acid rain may be deposited in water, on vegetation, and on soils, creating acidic conditions with adverse effects on flora and fauna. Ultimately, acidification affects the ability of ecosystems to provide ecosystem services, such as carbon and nutrient cycling, and support for plant, animal, and human life. We will discuss the Acid Rain Program of the CAA below.

Research has demonstrated that increased ground-level ozone causes damage to cell membranes on plants and inhibits key processes required for their growth and development. The loss of plant cover affects many ecosystems. Trees and other vegetation absorb pollutants such as excess nitrogen dioxide, ozone, and particulate matter through their leaves and needles. By reducing these pollutants in the air, plants help to improve air quality. Decreased plant cover means less filtering capacity to clean the air.[2]

Eutrophication, the process of accumulation of nutrients, including nitrogen, in water bodies, often results from air pollution. Nutrient overloads in aquatic ecosystems can cause algae blooms and ultimately a loss of oxygen and a loss of aquatic life. As ecosystems are impacted, so is biological diversity.

Ultimately, human populations are affected along with natural ecosystems. Harmful concentrations of pollutants may enter drinking water, notably through groundwater seepage. In addition, water quality may deteriorate as air pollution negatively affects vegetation, which offers natural filtration of water systems used by people. Affected vegetation also loses much of its ability to capture carbon and reduce the impacts of climate change.

Air pollution by chlorofluorocarbons (CFCs), used for years as refrigerants, has been associated with a depletion of atmospheric ozone. Atmospheric ozone is important in intercepting dangerous ionizing radiation from outside the atmosphere, which would otherwise damage tissues and even the DNA of living things. With the buildup of CFCs in the global atmosphere, there has been a measurable loss of ozone, especially near the poles. This has resulted in increases in ionizing radiation from outside our atmosphere and measurable damage to reproduction and growth in plants and animals. However, there has been a worldwide effort to reduce CFC use over the past several decades, and there is evidence that atmospheric ozone is returning as CFC pollution has decreased.

6.5 THE REGULATION OF AIR POLLUTION: THE CLEAN AIR ACT

Air pollution has been recognized by human cultures for millennia. Air quality issues were found in any large city where people burned wood and worked crafts or industry, whether or not they used fossil fuels. In first-century Rome, air pollution was known as *gravioris caeli* (heavy heaven) or *infamis aer* (infamous air).

Air pollution was recognized in Europe as early as the industrial revolution when the burning of fossil fuels created deadly conditions combined with congested living conditions. The first "smoke abatement" ordinances were used to forbid coal burning in London in 1273 during the reign of Edward I of England. These early ordinances were used to convict (and execute) violators.

Similar circumstances occurred in the United States during the nineteenth century and led to the first smoke abatement ordinances in several larger, industrial U.S. cities (e.g., Chicago and Cincinnati). These early ordinances required furnaces to consume their own smoke or forbade burning high-sulfur coal, and levied fines of $10–100 for offenders. These early smoke abatement ordinances were based on common law nuisance (public or private). Despite challenges by a regulated community that was used to getting its own way, these early controls were held by the U.S. Supreme Court to be a valid exercise of the "police power" of the municipality to protect public health and welfare. However, these early ordinances based on common law were problematic because of the difficulty of proving actual harm, and difficulties in establishing and measuring ambient concentrations of pollutants.

Today, there are several federal laws that affect air quality, but the most significant by far is the federal CAA. As we will see, its provisions have contributed significantly to air quality improvements in the United States, and in other parts of the world as well.

What Is Air Pollution?

Before we address the CAA specifically, let's determine what we mean when we talk about air pollution. *Air pollution* is defined as substances or particles that are undesirable in ambient (outdoor) air. Some air pollution occurs naturally, such as when the decay of natural organic materials releases hydrocarbons and H_2S (hydrogen sulfide) into the air. The air pollutants of greatest concern are anthropogenic (man-made) pollutants, and the effects they have on both ecosystems and human health and welfare. Table 6.4 lists some important air pollutants, their sources, and health effects. Table 6.4

There are, of course, many pollutants that may reach the ambient air that we breathe. The CAA currently recognizes six "criteria pollutants" that receive special attention. These six criteria pollutants were selected because their health effects are known, and relatively simple technology is available to measure them. The list includes carbon monoxide (CO), ground-level ozone (O_3), particulate matter, nitrogen dioxide (NO_2), and sulfur dioxide (SO_2). The original list included "hydrocarbons," but these were dropped because causation of health effects was extremely difficult to prove. However, "lead" (Pb) was subsequently added following several court decisions. Note that carbon dioxide (CO_2) is not a criteria pollutant under the CAA.

Table 6.4 Common Ambient (Outdoor) Air Pollutants, Pollutant Sources and Their Effects on Health (* Are Criteria Air Pollutants under the Clean Air Act)

Ambient Air Pollutant	*Air Pollutant Sources*	*Health Issues*
*Carbon monoxide (CO)	Burning fossil fuels and wood	Increases confusion, sleepiness, low blood oxygen level, slow reflexes
Carbon dioxide (CO_2)	Burning fossil fuels	Lowers oxygen levels, causes vision defects, reduces respiratory and brain functions.
*Nitrogen dioxide (NO_2)	Burning fuels, electricity generation, vehicle engines	Causes lung dysfunction and can cause bronchitis in asthmatic children. Potentially toxic.
*Sulphur dioxide (SO_2)	Burning fossil fuels and industrial processes	Eye irritation and respiratory inflammation, asthma, mucus secretion, decreases pulmonary function.
*Ozone (O_3)	Photochemical smog produced by the interaction of sunlight and air pollutants (mainly NO_3)	Breathing difficulties and asthma, and eye and lung irritation.
*Suspended particulate matter (PM_{10}, $PM_{2.5}$)	Mixture of solid and liquid organic plus inorganic materials, mineral dust, and water	Disrupts gas exchange in the lungs and causes respiratory illness.
*Lead (Pb)	Burning leaded gasoline, ore, and metal smelting	Impairs the nervous system, kidney function, immune system, reproductive, and cardiovascular system.

After Kapoor, M. 2017. Managing Ambient Air Quality Using Ornamental Plants-An Alternative Approach. Universal Journal of Plant Science. 5, 1–9. 10.13189/ujps.2017.050101.

The Clean Air Act

Most of the basic structure of the Clean Air Act (CAA) was established by Congress in 1970. There have been periodic amendments to the original CAA, particularly in 1977 and 1990.

The structure of the CAA is based on a series of federal/state partnerships. The federal EPA sets air quality and emission standards, but states determine how the federal standards are met and issue permits to sources of air pollution. The CAA recognizes that stationary and mobile air pollution sources should be regulated separately. In general, Title I deals with stationary sources, Title II with mobile sources, and Title III with administration and judicial review procedures.

A centerpiece of the CAA is the requirement that each state develop a State Implementation Plan (SIP) that explains how the state will attain the NAAQS by the applicable deadlines. SIPs must be approved by the EPA as containing sufficient measures to attain NAAQS and meet other requirements described in the CAA. SIPs must contain air pollution measures in adopted, "regulatory" form within one year after approval by the EPA. Upon approval by the EPA, SIP requirements can be enforced against regulated sources by the EPA as well as by any citizen.

For purposes of State Implementation Plans, all Air Quality Control Regions are divided into *attainment areas* (those areas that have met the NAAQS for a particular criteria pollutant) and *nonattainment areas* (those areas that have not met the NAAQS for a criteria air pollutant). Of greatest interest to us are nonattainment areas, which have failed to meet primary or secondary NAAQS for a criteria pollutant.

Performance standards in nonattainment areas require that each existing stationary source (excluding new or mobile sources) meet the standard of Reasonably Available Control Technology (RACT). EPA regulations define RACT as devices, systems, process modifications, etc. that permit attainment of limits for stationary sources established by EPA.[3] This RACT standard can be considered a "baseline" standard. Each *new* stationary source (or modification of an existing stationary source) must obtain a permit and demonstrate that it will use the Lowest Achievable Emission Rate (LAER) for each nonattainment pollutant. This LAER standard is far stricter than the RACT standard. The idea is that it is cheaper and more efficient to build more advanced (but more expensive) air pollution controls at the time a facility is built or modified than it is to retrofit an existing facility. This concept is called "technology forcing," and it is used frequently in environmental laws and regulations. The procedures and equipment used to reduce air pollution from stationary sources include a variety of filters, centrifugal collection, and electrostatic and chemical treatments.

Clean Air Act Title I

Title I of the CAA requires EPA to set up three different kinds of nationwide standards. First, EPA created Air Quality Control Regions and developed NAAQS, which determine the maximum concentrations of designated "criteria" pollutants in ambient air. Second, EPA designated New Source Performance Standards (NSPS), which establish allowable emission limitations for various kinds of new stationary sources of air pollution. Third, EPA must set National Emissions Standards for Hazardous Air Pollutants (NESHAP) for which no ambient air quality standards exist.

The CAA defines an air quality control region (AQCR) as a contiguous area where air quality, and thus air pollution, is relatively uniform. In those cases where topography is a factor in air movement, AQCRs often correspond with airsheds. AQCRs for the entire United States are now in place for each of the six criteria pollutants.

The CAA requires the EPA to prepare NAAQSs for criteria pollutants that have an adverse effect on public health and welfare. "Primary standards" were developed that are, in essence, health-based standards. "Secondary standards" were developed that are necessary to protect the public welfare, so they are welfare-based standards.

Primary and secondary NAAQS standards for each of the criteria pollutants must "reflect the latest scientific knowledge useful in indicating the kinds and extent of all identifiable effects on public health and welfare."[4] These standards must establish a minimally acceptable level of ambient air quality that protects humans and the human environment from all known effects, plus any unknown effects that are legitimate sources of concern. Current primary and secondary standards for the six criteria pollutants can be found on EPA's website.[5]

Title I also requires that EPA establish NESHAP for which no ambient air quality standards exist. "Hazardous" air pollutants (HAPs, or "air toxics") are defined as any of the 189 pollutants listed in CAA §112:[6] Congress established the original list of 189 hazardous air pollutants, although EPA (or a petitioner) can add to the list any substances it finds pose a "threat of adverse human health effects" or environmental effects, including any toxic, carcinogenic, mutagenic, teratogenic, or neurotoxic substance. It is sufficient for listing purposes that a *possibility* exists that a substance poses a health threat.

EPA is required to publish a list of area source categories of listed hazardous air pollutants. This list must include enough source categories to ensure that 90 percent of the area source emissions of the thirty hazardous pollutants posing the greatest threat to public health in the most cities are subject to §112 regulation. Under CAA §112(a)(1), a "major source" is any stationary source (or combination of sources) that produce 10 tons/year of any hazardous air pollutant, or 25 tons/year of any combination of hazardous pollutants. An "area source" is any source of hazardous air pollutants that is not a major source. EPA is in the process of establishing special regulatory programs that apply to major and area sources, under which major sources will have special limitations that do not apply to area sources. Obviously, a source would rather be regulated as an "area source" than as a "major source."

Technology-based emission standards for hazardous pollutants must require the maximum emissions reduction that the EPA determines to be achievable, including a prohibition on any emission (MACT). For a new source, emission standards must be at least as stringent as the emission control achieved by the best-controlled similar source. Together, these standards are designed to minimize the health and environmental dangers posed by these dangerous air pollutants.

The Acid Rain Program

Acid rain was first identified as an environmental issue in the 1970s. The northeastern United States was especially hard hit. In 1980, the National Academy of Sciences issued a report suggesting that a major culprit was emissions of sulfur dioxide (SO_2) from electricity-generating power plants in the Midwest that were being carried eastward by prevailing winds. The emissions were then transformed in the atmosphere into sulfuric acid (H_2SO_4), which then fell to the surface of the earth as acidic precipitation, or "acid rain." It has subsequently been suggested that both SO_2 and NO_x can form acid rain (figure 6.8).

Acid rain causes the acidification of lakes and streams, with subsequent destruction of plants and wildlife. Acid rain contributes to the damage of trees at high elevations (for example, red spruce trees above 2,000 feet in elevation). In addition, acid rain accelerates the decay of building materials and paints, including irreplaceable buildings and artistic structures such as

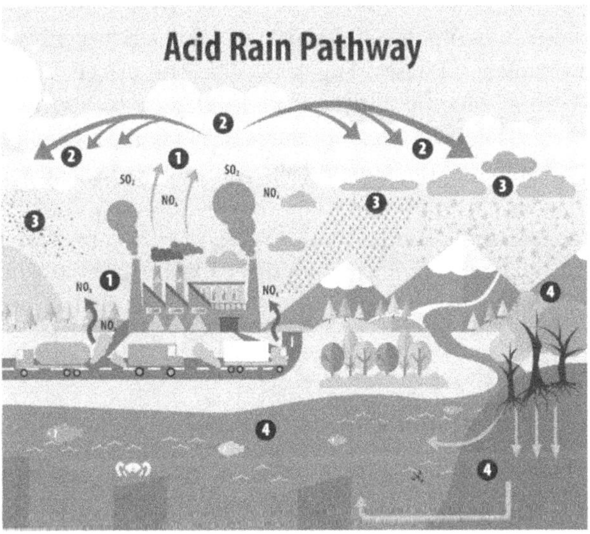

Figure 6.8 Acid Rain Formation. This image illustrates the pathway for acid rain in our environment: (1) Emissions of SO_2 and NO_x are released into the air, where (2) the pollutants are transformed into acid particles that may be transported long distances. (3) These acid particles then fall to the earth as wet and dry deposition (dust, rain, snow, etc.) and (4) may cause harmful effects on soil, forests, streams, and lakes. *Source*: EPA, https://www.epa.gov/acidrain/what-acid-rain

statues and sculptures. Prior to falling to the earth, SO_2 and NO_x gases and their particulate matter derivatives, sulfates and nitrates, contribute to visibility degradation and impact public health.

The primary source of SO_2 pollution is coal-burning power plants. SO_2 is created when the sulfur in coal is released during combustion and reacts with oxygen in the air. The amount of sulfur dioxide created depends on the amount of sulfur in the coal. While all coal contains some sulfur, the amount varies significantly depending on where the coal is mined. Atmospheric NO_x is from coal and/or oil-burning plants, and nitrogen oxides from car and truck exhaust and from oil-burning plants.

Congress responded to concerns about acid rain with the Acid Precipitation Act of 1980, which authorizes the EPA to develop an Acid Rain Program called the National Acid Precipitation Abatement Program (NAPAP). The overall goal of the Acid Rain Program is to achieve significant environmental and public health benefits through reductions in emissions of SO_2 and NO_x. To achieve this goal at the lowest cost to society, the program employs both traditional and innovative, market-based approaches for controlling air pollution. In addition, the program encourages energy efficiency and pollution prevention.

Title IV of the CAA sets as its primary goal the reduction of annual SO_2 emissions by ten million tons below 1980 levels. To achieve these reductions, the CAA requires a two-phase tightening of the restrictions placed on fossil fuel-fired power plants. Phase I began in 1995 and affects 263 units at 110 mostly coal-burning electric utility plants located in twenty-one eastern and midwestern states. Phase II began in the year 2000 and tightened the annual emissions limits imposed on large, higher-emitting plants, and set restrictions on smaller, cleaner plants fired by coal, oil, and gas.

There is considerable evidence that the NAPAP program is working. We have seen reductions from many power plants, and the acidification of environments, especially in the northeastern United States, appears to be decreasing.

Clean Air Act Title II and Mobile Sources

CAA Title II deals with mobile sources, which have become a frequent source of lawsuits and political posturing. The 1970 CAA required the EPA to adopt emission limitations for motor vehicles to achieve reductions in emissions from motor vehicles, as well as from other mobile sources such as locomotives, airplanes, and ships. States are preempted from adopting emission limitations for motor vehicles and certain other mobile sources, although under CAA §177, there is an exception for the state of California.[7] California is free to adopt its own (more stringent) motor vehicle standards, and standards for some (but not all) other mobile sources. Other states can adopt the California standards if they wish.

Title II mandates that improved gasoline formulations be sold in some polluted cities to reduce emissions of carbon monoxide or ozone-forming hydrocarbons. Other programs set low vehicle emission standards to stimulate the introduction of cleaner cars and fuels. In cities that are in ozone nonattainment areas, growth in vehicle travel must be limited by encouraging alternatives to solo driving.

Other Title II mandates include automobile pollution control devices (catalytic converters) and vapor recovery nozzles on gas pumps in ozone nonattainment areas. The 1990 amendments required that oxygenated gasoline (which reduces emissions of carbon monoxide) be sold in the worst carbon monoxide (CO) nonattainment areas, and that "reformulated" gasoline (RFG), designed to reduce emissions of volatile organic compounds and toxic air pollutants, be sold in the ten worst ozone nonattainment areas. Other ozone nonattainment areas can opt into the RFG program.

Alternative fuels and cleaner engines are stimulated by several programs. Under CAA § 209(b), California can develop emission standards more stringent than the federal standard, and other states can then adopt the California standard. California has developed a program requiring low emission vehicles (LEVs), ultralow emission vehicles (ULEVs), and zero emission vehicles (ZEVs), and several northeastern states have adopted similar requirements. Other states may choose to follow California's lead.

6.6 THE FUTURE OF THE GLOBAL ATMOSPHERE

As we have seen, our atmosphere is a dynamic and critically important part of the global environment. But it is under a series of existential threats, both anthropogenic and natural. We have seen that the legal framework exists to prevent the worst of atmospheric degradation, but there is evidence that it may not be enough to prevent environmental catastrophe. We will explore these issues and climate change in much greater detail in chapter 12.

Before we leave our discussion of air, atmosphere, and climate, it is important to recognize how much environmental progress we have made as a nation in the past 60 years. As recently as the 1950s and 1960s, most large U.S. cities suffered from serious, often deadly, air pollution. There are photographs in the Carnegie Museum from the first half of the twentieth century showing downtown Pittsburgh, PA, at midday with air pollution so bad that the automatic streetlights came on. At the time, it was easy to believe that air pollution would only get worse. However, since the introduction of the CAA and other environmental laws in the 1960s and 1970s, there have been dramatic improvements in air quality across much of the United States. The point of this observation is that there is hope for the future. We will discuss more about our global atmospheric future in chapter 12, our chapter on climate change.

* * *

Review Questions

1. List the five layers of the atmosphere and the characteristics of each.
2. What are the major meteorologically important gases, and how do they contribute to weather?
3. Describe how uneven heating of the earth forms weather.
4. What causes El Niño (Southern Oscillation) weather events, and what changes occur?
5. Explain why deserts typically form on the downwind sides of mountain ranges.
6. Distinguish between sulfurous (London) smog and photochemical smog.
7. How does acid rain form, and what aspects of the Clean Air Act react to the problem?
8. What are criteria pollutants under the CAA, and why were they selected?
9. What are National Ambient Air Quality Standards (NAAQS) under the CAA?
10. Describe how the CAA regulates mobile sources of air pollution.

Case Analysis

As we learned, Section 112 of the CAA requires the EPA to regulate hazardous air pollutants (HAPs). The EPA may regulate power plants only if it concludes that "regulation is appropriate and necessary" after studying hazards to public health. In 2000, the EPA found power-plant regulation to be "appropriate" because power plant emissions pose risks to public health and the environment and because controls capable of reducing these emissions were available. It found regulation "necessary" because other CAA requirements did not eliminate those risks. The EPA estimated that the cost of power plant regulation would be $9.6 billion a year, but that quantifiable benefits from the reduction in hazardous-air-pollutant emissions would be $4-$6 million a year. In other words, the EPA refused to consider the effect of costs to the power plants in setting HAPs emission standards. Several states sued, and the case eventually reached the U.S. Supreme Court. The following is an abstract of the 2015 case *Michigan v. EPA*.

> The Clean Air Act directs the Environmental Protection Agency to regulate emissions of hazardous air pollutants from power plants if the Agency finds regulation "appropriate and necessary." We must decide whether it was reasonable for EPA to refuse to consider cost when making this finding.
>
> . . .
>
> The Clean Air Act establishes a series of regulatory programs to control air pollution from stationary sources (such as refineries and factories) and moving sources (such as cars and airplanes). 69Stat. 322, as amended, 42 U. S. C. §§7401–7671q. One of these is the National Emissions Standards for Hazardous Air Pollutants Program—the hazardous-air-pollutants program, for short. Established in its current form by the Clean Air Act Amendments of 1990, this program targets for regulation stationary-source emissions of more than 180 specified "hazardous air pollutants." §7412(b).
>
> For stationary sources in general, the applicability of the program depends in part on how much pollution the source emits. A source that emits more than 10 tons of a single pollutant or more than 25 tons of a combination of pollutants per year is called a major source. §7412(a)(1). EPA is required to regulate all major sources under the program. §7412(c)(1)–(2). A source whose emissions do not cross the just-mentioned thresholds is called an area source. §7412(a)(2). The Agency is required to regulate an area source under the program if it "presents a threat of adverse effects to human health or the environment . . . warranting regulation." §7412(c)(3).
>
> . . .
>
> EPA completed the study required by §7412(n)(1)(A) in 1998, and concluded that regulation of coal- and oil-fired power plants was "appropriate and necessary" in 2000. In 2012, it reaffirmed the appropriate-and-necessary finding, divided power plants into subcategories, and promulgated floor standards. The Agency found regulation "appropriate" because (1) power plants' emissions of mercury and other hazardous air pollutants posed risks to human health and the environment and (2) controls were available to reduce these emissions. It found regulation "necessary" because the imposition of the Act's other requirements did not eliminate these risks. Ibid. EPA concluded that "costs should not be considered" when deciding whether power plants should be regulated under §7412. Id., at 9326.
>
> In accordance with Executive Order, the Agency issued a "Regulatory Impact Analysis" alongside its regulation. This analysis estimated that the regulation would force power plants to bear costs of $9.6 billion per year. The Agency could not fully quantify the benefits of reducing power plants' emissions of hazardous air pollutants; to the extent it could, it estimated that these benefits were worth $4 to $6 million per year. Ibid. The costs to power plants were thus between 1,600 and 2,400 times as great as the quantifiable benefits from reduced emissions of hazardous air pollutants. The Agency continued that its regulations

would have ancillary benefits—including cutting power plants' emissions of particulate matter and sulfur dioxide, substances that are not covered by the hazardous-air-pollutants program. Although the Agency's appropriate-and-necessary finding did not rest on these ancillary effects, id., at 9320, the regulatory impact analysis took them into account, increasing the Agency's estimate of the quantifiable benefits of its regulation to $37 to $90 billion per year, id., at 9306. EPA concedes that the regulatory impact analysis "played no role" in its appropriate-and-necessary finding.

. . .

Petitioners (who include 23 States) sought review of EPA's rule in the Court of Appeals for the D. C. Circuit. As relevant here, they challenged the Agency's refusal to consider cost when deciding whether to regulate power plants. The Court of Appeals upheld the Agency's decision not to consider cost (in 2014).

. . .

EPA's decision to regulate power plants under §7412 allowed the Agency to reduce power plants' emissions of hazardous air pollutants and thus to improve public health and the environment. But the decision also ultimately cost power plants, according to the Agency's own estimate, nearly $10 billion a year. EPA refused to consider whether the costs of its decision outweighed the benefits. The Agency gave cost no thought at all, because it considered cost irrelevant to its initial decision to regulate.

EPA's disregard of cost rested on its interpretation of §7412(n)(1)(A), which, to repeat, directs the Agency to regulate power plants if it "finds such regulation is appropriate and necessary." The Agency accepts that it could have interpreted this provision to mean that cost is relevant to the decision to add power plants to the program. Tr. of Oral Arg. 44. But it chose to read the statute to mean that cost makes no difference to the initial decision to regulate. See 76 Fed. Reg. 24988 (2011) ("We further interpret the term 'appropriate' to not allow for the consideration of costs").

. . .

The Clean Air Act treats power plants differently from other sources for purposes of the hazardous-air-pollutants program. Elsewhere in §7412, Congress established cabined criteria for EPA to apply when deciding whether to include sources in the program. It required the Agency to regulate sources whose emissions exceed specified numerical thresholds (major sources). It also required the Agency to regulate sources whose emissions fall short of these thresholds (area sources) if they "presen[t] a threat of adverse effects to human health or the environment . . . warranting regulation." §7412(c)(3). In stark contrast, Congress instructed EPA to add power plants to the program if (but only if) the Agency finds regulation "appropriate and necessary." §7412(n)(1)(A). One does not need to open up a dictionary in order to realize the capaciousness of this phrase. In particular, "appropriate" is "the classic broad and all-encompassing term that naturally and traditionally includes consideration of all the relevant factors." Although this term leaves agencies with flexibility, an agency may not "entirely fai[l] to consider an important aspect of the problem" when deciding whether regulation is appropriate.

Read naturally in the present context, the phrase "appropriate and necessary" requires at least some attention to cost. One would not say that it is even rational, never mind "appropriate," to impose billions of dollars in economic costs in return for a few dollars in health or environmental benefits. In addition, "cost" includes more than the expense of complying with regulations; any disadvantage could be termed a cost. EPA's interpretation precludes the Agency from considering any type of cost—including, for instance, harms that regulation might do to human health or the environment. The Government concedes that if the Agency were to find that emissions from power plants do damage to human health, but that the technologies needed to eliminate these emissions do even more damage to human health, it would still deem regulation appropriate.

. . .

We hold that EPA interpreted §7412(n)(1)(A) unreasonably when it deemed cost irrelevant to the decision to regulate power plants. We reverse the judgment of the Court of Appeals for the D. C. Circuit and remand the cases for further proceedings consistent with this opinion.[8]

Questions

1. Both the Appellate Court and the Supreme Court said that EPA "refused" to consider cost to the power plants. Is that correct? What did EPA find that the cost would be relative to the benefits?
2. The Supreme Court talks a lot about the dictionary meanings of the words "appropriate" and "necessary." Is this the correct way to interpret what Congress meant in the Clean Air Act? What might be another way to determine the meanings of the words in the statute?
3. Did the Supreme Court consider the EPA's determination regarding the urgency of the regulations? What is the effect of the Supreme Court's decision on public health and welfare?
4. In this and many other environmental law cases, the Supreme Court does not comment on the effects of regulations on health and welfare. Is it possible that this will change with the 2024 Supreme Court decision in *Loper Bright Enterprises v. Raimondo*,[9] which held that courts, not federal agencies, must interpret the meanings of federal laws?
5. Did the Supreme Court get this right? Should the Court consider the effects on health and welfare in decisions like this? If you were making the decision, how would you rule?

NOTES

1. 40 CFR §50.1(e).

2. For a review of scientific information on air pollution and ecosystems, see: Lovett, G. M., Tear, T. H., Evers, D. C., Findlay, S. E., Cosby, B. J., Dunscomb, J. K., Driscoll, C. T., & Weathers, K. C. 2009. Effects of Air Pollution on Ecosystems and Biological Diversity in the Eastern United States. *Annals of the New York Academy of Sciences, 1162*(1), 99–135.

3. 40 CFR §51.1(o).

4. 42 U.S.C. §7408(a)(2).

5. https://www.epa.gov/naaqs.

6. 42 U.S.C. §7412.

7. 42 U.S.C §7507.

8. *Michigan v. Environmental Protection Agency*, 576 U.S. 743, 135 S.Ct. 2699 (2015).

9. 603 U.S. ___ (2024).

Chapter 7

Aquatic Ecosystems

Chapter Learning Outcomes

At the end of this chapter, the reader will be able to:

1. *Explain why Earth is a "water planet."*
2. *Contrast the composition and the organisms of a rocky and a sandy coastline ecosystem.*
3. *Describe the structure of coral ecosystems.*
4. *List the important characteristics of saltwater marshes and mangrove forest ecosystems.*
5. *List and describe the five layers of open ocean ecosystems.*
6. *Explain the unique characteristics of hydrothermal vent communities.*
7. *Describe the characteristics of the five zones of a freshwater lake.*
8. *Contrast oligotrophic and eutrophic lakes.*
9. *Contrast the composition of the three zones of a river ecosystem.*
10. *Contrast the composition and the organisms of a freshwater marsh and a freshwater swamp.*
11. *Explain the significance of aquatic transition zones like estuaries.*

The next step in our investigation of the "media" of the environment (air, water, and land) is an investigation of aquatic ecosystems. We begin with an introduction to the ecology of marine (ocean), freshwater, and transition ecosystems. Then we will investigate the regulatory system that supports these valuable ecosystems and regulates water pollution in its various forms. Finally, we will examine the special significance of marine and freshwater wetlands, and the legal system that protects them.

7.1 EARTH IS A WATER PLANET

It is completely reasonable to characterize the earth as a "water planet."[1] In fact, water forms over 75 percent of the earth's surface. Water is a critical component of life on earth and the support of ecosystems of all types and is an important driver of the formation of landforms. We will explore the significance of water throughout this book.

An interesting way to view the surface of the earth is to present it as an ocean hemisphere and an ocean-land hemisphere, rather than the more familiar eastern and western hemispheres we know from geography books. Figures 7.1a and 7.1b shows the ocean and the ocean-land hemispheres. In this view, the ocean hemisphere, centered on the southern Pacific Ocean, is approximately 90 percent water. The ocean-land hemisphere, centered on the northern Atlantic and Arctic Oceans, is about 50 percent water.

Figure 7.1 Earth Is a Water Planet. This presentation demonstrates the ocean hemisphere (left) and the land–ocean hemisphere (right). Ocean Hemisphere

Figure 7.1b Earth Is a Water Planet. This presentation demonstrates the ocean hemisphere (left) and the land–ocean hemisphere (right). Ocean-Land Hemisphere. *Source*: By Citynoise - Own work, CC BY-SA 4.0, https://commons.wikimedia.org/w/index.php?curid=42626963

7.2 INTRODUCTION TO AQUATIC ECOSYSTEMS

Most of the earth's water is in its interconnected oceans, which cover almost 90 percent of the planet's mostly ocean hemisphere and half of its ocean-land hemisphere. Freshwater systems cover less than 2.5 percent of the earth's surface.

We will examine marine and freshwater ecosystems and explore the transition zones where they come into contact.

7.3 MARINE (SALTWATER) ECOSYSTEMS

Marine ecosystems are aquatic environments with relatively high levels of dissolved salt (NaCl), in contrast with freshwater ecosystems, which have a significantly lower salt content. Marine waters cover more than 70 percent of the surface of the earth and account for more than 97 percent of earth's water supply and 90 percent of habitable space on earth. Seawater has an average salinity of thirty-five parts per thousand (0/00) of water, although actual salinity varies among marine ecosystems.

Marine ecosystems are important sources of food and jobs for a significant portion of the global human population. Unfortunately, human overuse and pollution are significant threats to marine ecosystems. Environmental damage to marine ecosystems includes the unsustainable exploitation of marine resources (for example, overfishing of certain species), marine pollution, climate change, and the development of coastal areas. In addition, much of the carbon dioxide causing global warming is absorbed by the ocean, and ocean chemistry is changing through processes like ocean acidification, which in turn threatens marine ecosystems. We will discuss these issues below.

Coasts and Shorelines

For most of us, our first experience with marine ecosystems involved a visit to the seashore. Whether it was a sandy beach or a rocky shoreline, we were observing the important point of contact between marine and terrestrial ecosystems. The particular type of coast we experienced depends on local conditions. The actions of waves, tides, and currents all help create coastlines.

There are approximately 620,000 kilometers (390,000 miles) of coastline on the earth. Coasts are important sources of natural ecosystems, which support much of the earth's biodiversity. On land, they harbor important ecosystems such as freshwater or marine wetlands (discussed below), which are important for bird populations and a variety of other terrestrial animals. In wave-protected areas, they harbor salt marshes, mangroves, or seagrasses, all of which can provide nursery habitats for a variety of fish, invertebrates, and other aquatic species.

Coasts and shorelines are of special economic importance to humans, too. According to an analysis by the United Nations, 44 percent of all humans live within 150 kilometre (93 mi) of the coast.[2] Because of their significance to society and high population concentrations, coasts are important for major parts of the global food and economic structures. Important human activities are concentrated in port cities. Coastal fisheries (commercial, recreational, and subsistence) and aquaculture are major economic activities that create jobs and food, especially protein, for human populations worldwide.

From an ecosystem perspective, an important part of all coastlines is the *intertidal zone*, or the area between the low and high tide lines (between mean high water and low water marks). During low tide, the intertidal zone is exposed to the atmosphere, whereas during high tide the intertidal zone is underwater. Intertidal ecologists study the interactions between intertidal organisms and their environment, as well as between different species of intertidal organisms within a particular intertidal community. The most important environmental and species interactions may vary based on the type of intertidal community and substrates.

A useful comparison of intertidal communities is between sandy and rocky coastlines. The major difference between rocky and sandy shore habitats is the nature of the substrate. Rocky coastlines tend to occur where there is higher wave energy that washes away sand particles. By contrast, sandy beaches generally have much lower wave energy, which allows sand to accumulate, often resulting in barrier island formation and inland sand dunes.

Rocky Coastlines

Rocky coastlines may have steep gradients and may have tide pools, boulder fields, or even corals in tropical areas. Biodiversity in these areas can be very high. Oceanographers divide rocky coasts into four zones: a spray zone, a high tide zone, a middle tide zone, and a low tide zone (figure 7.2).

Ecologists discovered many years ago that rocky shore organisms tend to occur in zones between the high tide mark and the low tide mark. Species closer to the high tide mark often have adaptations to survive exposure to air and sunlight without drying out, but at the expense of relatively high predation. Species closer to the low tide mark are better at avoiding predation but are more susceptible to injury during long periods out of the water. Rocky shore organisms differ depending on the geographical characteristics of the area, but commonly include barnacles, mussels, limpets, sea anemones, and sea stars, each with a different ability to avoid predation or survive outside of the water. Figure 7.2 shows the distributions of some of these organisms.

Figure 7.2 Some Characteristic Organisms of a Rocky Shoreline. *Source*: Image by brgfx</a> on Freepik. Rocky shore zonation education diagram. <a href="https://www.freepik.com/free-vector/rocky-shore-zonation-education-diagram_37530971.htm">

Sandy Beaches

Sandy beaches are loose deposits of sand, which often include gravel and shells. They make up a large portion (about 30 percent) of the world's ice-free coastlines. Sandy beaches serve as buffer zones that protect the coastline, sea cliffs, or dunes from direct wave action. It is an extremely dynamic environment where sand, water, and air interact. Most beaches have a gentle slope, although fine-grained sand beaches tend to be nearly flat.

Oceanographers divide sandy beaches into an upper *supratidal zone* that often includes sand dunes and beach grasses, an *upper, middle, and lower intertidal zone*, and a *subtidal zone* (permanently submerged). These zones are shown in figure 7.3.

Organisms of sandy beaches are divided into *macrofauna* (>1 mm) and smaller *meiofauna*. The macrofauna of sandy beaches includes most major invertebrate taxa and is highly variable, although crustaceans and insects dominate the upper intertidal zone, polychaete worms in the middle zone, and mollusks in the lower intertidal zone (figure 7.3). Of course, larger organisms like birds are also part of the macrofauna.

The organisms of the meiofauna inhabit the interstitial spaces between the highly porous sand grains. These organisms are unseen but may be extremely abundant and an important part of beach ecosystems, especially as a food source for the macrofauna. The dominant taxa of sandy beach meiofauna

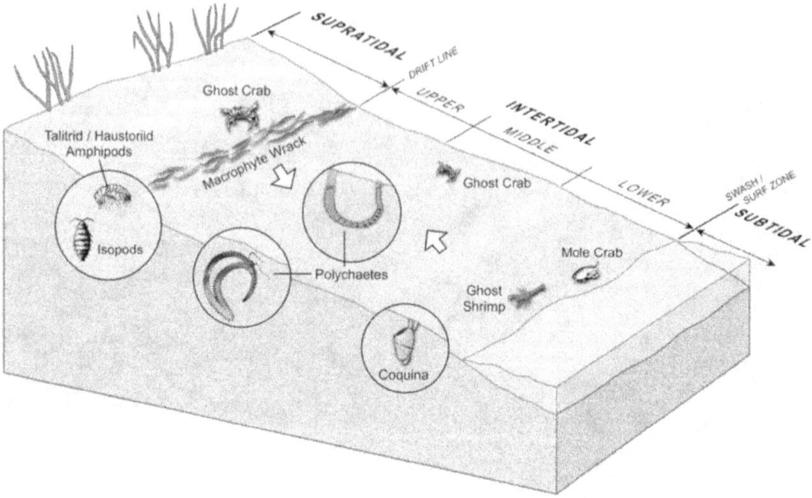

Figure 7.3 Some Characteristic Organisms of a Sandy Beach. *Source*: Michel J, Fegley SR, Dahlin JA, Wood C (2017). Oil spill response-related injuries on sand beaches: When shoreline treatment extends the impacts beyond the oil. Mar Ecol Prog Ser 576:203–218. https://doi.org/10.3354/meps11917

are tiny nematodes (roundworms), copepods (crustaceans), and other groups including turbellarians and oligochaetes (segmented worms).

There are other intertidal communities such as *salt marshes*, *mangrove swamps*, and *coral reefs*, which we will explore later in this chapter.

Coral Reefs

Coral reefs are probably the most iconic of all coastal ecosystems. They are famous for their unrivaled beauty and waters teeming with life. In fact, coral reefs have some of the greatest biodiversity of any ecosystem in the world. Their distribution is mostly in tropical and semitropical waters, between 30° north and 30° south latitudes (figure 7.4).

Generally, there are about twice as many coral species in Pacific Ocean reefs as in Atlantic Ocean reefs. In fact, the most diverse ecosystem in the world is the huge Coral Triangle in Southeast Asia. The Coral Triangle extends from the Philippines in the north to the Solomon Islands in the east and to the islands of Indonesia and Papua in the west.

Coral reefs are formed by colonies of coral *polyps*, the tiny animals primarily responsible for building reefs. These polyps deposit the hard nonliving "skeleton" of the reef, composed primarily of calcium carbonate ($CaCO_3$). Polyps can take many forms, from large reef-building colonies and graceful flowing fans to small, solitary organisms. There are thousands of species of

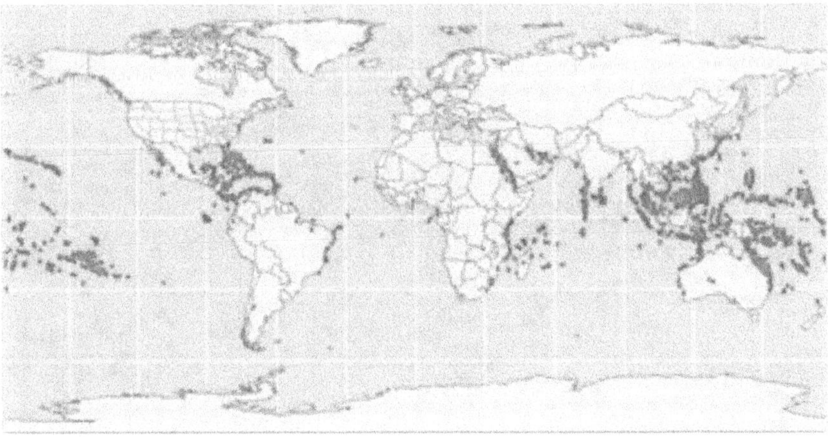

Figure 7.4 World Map of Major Coral Reef Sites. Coral reefs are seen as red dots on this world map. *Source*: NOAA Coral Reef Information System. https://www.coris.noaa.gov/about/what_are/

corals, some live in warm, shallow, tropical seas, and others in the cold, dark depths of the ocean.

Shallow-water, reef-building corals have a symbiotic relationship with photosynthetic algae called *zooxanthellae*, which live in their tissues. The coral provides a protective environment and the compounds zooxanthellae need for photosynthesis. In return, the algae produce oxygen and carbohydrates that the coral uses for food. The algae also help the coral remove waste. Because both partners benefit from the association, this type of symbiosis is called *mutualism*.

Deep-sea corals live in much deeper or colder oceanic waters and lack zooxanthellae. Unlike their shallow-water relatives, which rely heavily on photosynthesis to produce food, deep-sea corals feed on plankton and organic matter for much of their energy requirements.

Because of the diversity of life found in the habitats created by corals, reefs are often called the "rainforests of the sea." About 25 percent of marine fish species are dependent on healthy coral reefs. Fishes and other organisms find shelter, food, places for reproduction, and conditions for young to develop in the many crevices and spaces formed by coral reefs. The Northwest Hawaiian Island coral reefs, which are part of the Papahānaumokuākea National Marine Monument, provide an example of the diversity of life associated with shallow-water reef ecosystems. This area supports more than 7,000 species of fishes, invertebrates, plants, sea turtles, birds, and marine mammals. Other extensive coral reefs, like the Great Barrier Reef off the northern coast of Australia, are equally biodiverse. On the other hand, deep-water reefs or mounds are less well known but also support a wide variety of sea life in an otherwise mostly barren world. Figure 7.5 demonstrates a diverse Pacific coral reef.

Figure 7.5 Coral Reef. Rapture Reef, French Frigate Shoals, Papahānaumokuākea National Marine Monument. *Source*: James Watt. https://www.noaa.gov/education/resource-collections/marine-life/coral-reef-ecosystems

Coral reefs have tremendous economic value. They help to protect coastlines from storms and erosion, provide jobs for local communities, and offer opportunities for recreation. They are a source of food and new medicines. Over half a billion people depend on reefs for food, income, and protection. Fishing, diving, and snorkeling on and near reefs add hundreds of millions of dollars to businesses worldwide. The net economic value of the earth's coral reefs is estimated to be tens of billions of U.S. dollars per year.[3] Coral reef ecosystems are culturally important to indigenous people around the world.

Unfortunately, coral reef ecosystems are under severe threat. Some threats are natural, such as diseases, predators, and storms. Other threats are caused by humans and include pollution, sedimentation, unsustainable fishing practices, and climate change, which is raising ocean temperatures and causing ocean acidification. We will discuss these issues further in our discussion of climate change in chapter 12.

Many of these threats can stress corals, leading to coral bleaching and possible death, while others cause physical damage to these delicate ecosystems. During the 2014–2017 coral bleaching event, unusually warm waters (partially associated with a strong El Niño) affected 70 percent of coral reef ecosystems worldwide. Some areas were hit particularly hard, like the Great Barrier Reef in Australia, where hundreds of miles of coral were bleached.

Saltwater Wetlands

Another important type of saltwater ecosystems are coastal saltwater wetlands. In general, wetlands are areas where water covers the soil or is present either at or near the surface of the soil all year or for varying periods during the year. Water saturation (*hydrology*) largely determines how the soil develops and the types of plant and animal communities living in and on the soil. Wetlands may support both aquatic and terrestrial species. The prolonged presence of water creates conditions that favor the growth of specially adapted plants (*hydrophytes*) and promote the development of characteristic wetland (hydric) soils.[4]

Coastal saltwater wetlands provide several important environmental benefits. They act as buffers to the ocean and provide important flood protection, especially during storms, and provide significant shoreline stabilization. They help to mitigate river and stream flow maintenance. And saltwater wetlands also provide habitat for fish and wildlife, including endangered species.

There are several types of wetlands, but the most familiar wetland types are *marshes* (dominated by grasses) and *swamps* (dominated by trees). Our discussion here is specifically about saltwater wetlands, but we will have more to say about freshwater wetlands below.

Saltwater Marshes

Salt marshes are coastal wetlands that are flooded and drained by tides several times every day. They consist of salt-tolerant grasses that grow in marshy soils composed of deep mud and peat. Since salt marshes are regularly submerged by tides and contain a lot of slowly decomposing plant and animal material, oxygen levels in the soil can be extremely low. These conditions give salt marshes their characteristic rotten-egg odor. Figure 7.6 shows a typical salt marsh.

Salt marshes are important guardians of coasts and help to maintain healthy fisheries, coastlines, and communities. They provide shelter, food, and nursery grounds for more than 75 percent of coastal fishery species, including shrimp, crab, and many finfish. Salt marshes also protect shorelines from erosion by creating a buffer against wave action and by trapping soils. In flood-prone areas, salt marshes reduce the flow of floodwater and absorb rainwater. Salt marshes filter runoff and excess nutrients, and help to maintain water quality in coastal bays, sounds, and estuaries. Salt marshes and other coastal wetlands also serve as carbon sinks.

Figure 7.6 Saltwater Marsh Dominated by Smooth Cordgrass (*Sporobolus*). This marsh in coastal South Carolina, USA at low tide exposing the saturated soils. These wetlands are among the most energetically rich ecosystems on Earth. *Source*: Photo by the author

The United States has experienced tremendous losses of freshwater and coastal wetlands since the early twentieth century, primarily from construction, development, and habitat loss. The U.S. Fish and Wildlife Service estimated that between 2004 and 2009 alone, wetlands in coastal watersheds declined by 360,720 acres or over 80,000 acres per year.[5]

Mangrove Forest Swamps

Mangrove swamps are coastal wetlands found in tropical and subtropical regions around the world. There are about eighty different species of mangroves, all of which grow in areas with low-oxygen soil, where slow-moving waters allow fine sediments to accumulate. These wetlands are often found in estuaries, where fresh water meets salt water, and are infamous for their impenetrable maze of woody vegetation. In North America, they are found from the southern tip of Florida along the Gulf Coast to Texas. Florida's southwest coast supports one of the largest mangrove swamps in the world. Figure 7.7 shows a red mangrove forest swamp.

Mangrove swamps can be recognized by their dense tangle of prop roots that make the trees appear to be standing on stilts above the water. This tangle of roots allows the trees to handle the daily rise and fall of tides. The roots

Figure 7.7 A Mangrove Forest Swamp. A red mangrove forest (Rhizophora mangle) in Everglades National Park (Florida, USA). The arching roots are "prop roots" which provide habitat for many animal species above and below the water. Photo by Michael Johannes Osland, PhD, USGS. *Source*: Public domain. https://www.usgs.gov/media/images/prop-roots-red-mangrove-trees-everglades-national-park

slow the movement of tidal waters, causing sediments to settle out of the water and build up the muddy bottom. Mangrove forests stabilize coasts by reducing erosion from storm surges, currents, waves, and tides. The intricate root system of mangroves also makes these forests attractive to fish and other organisms seeking food and shelter from predators.

Mangrove swamps have extremely high biodiversity. Since these swamps are constantly replenished with nutrients transported by freshwater runoff from the land and flushed by the ebb and flow of the tides, they support a bursting population of bacteria, other decomposers and filter feeders. These ecosystems sustain billions of worms, protozoa, barnacles (*Balanus* spp.), oysters (*Crassostrea* spp.), and other invertebrates. These organisms, in turn feed fish and shrimp, which support wading birds, pelicans, and, famously, crocodiles.

As with other coastal ecosystems, there are worldwide threats to mangrove swamps. Changes in sea level would alter flooding patterns and the structure and areal extent of mangroves. Climate change may also alter rainfall patterns, which would, in turn change local salinity regimes and competition between mangroves and other wetland species. Increases in frequency or intensity of tropical storms and hurricanes, in combination with sea-level rise, may alter erosion and sedimentation rates in mangrove forests. Major local threats to mangrove ecosystems worldwide include clear-cutting and trimming of forests for urban, agricultural, or industrial expansion, as well as hydrological alterations, toxic chemical spills, and eutrophication.[6]

Oceans

The ocean or *oceanic zone* is the area of the ocean lying beyond the continental shelf (e.g., the *neritic zone*), or the area where water depths drop to below 200 meters (660 ft), seaward from the coast into the open ocean with its *pelagic zone*. Open oceans cover nearly three-quarters of the surface of the earth, so it is not surprising that they are responsible for much of the earth's biodiversity.

The oceanic zone has a variety of undersea terrains, including deep-sea volcanoes and basins, and deep trenches that are often deeper than Mount Everest is tall. While it is often difficult for life to sustain itself in this type of environment, many species have adapted and thrive in the oceanic zone, even in some of the deepest places.

The economic importance of the earth's oceans is huge. In 2022, the United Nations estimated that trade in ocean-based goods and services is worth at least $2.5 trillion per year, or about 3 percent of global GDP in 2020.

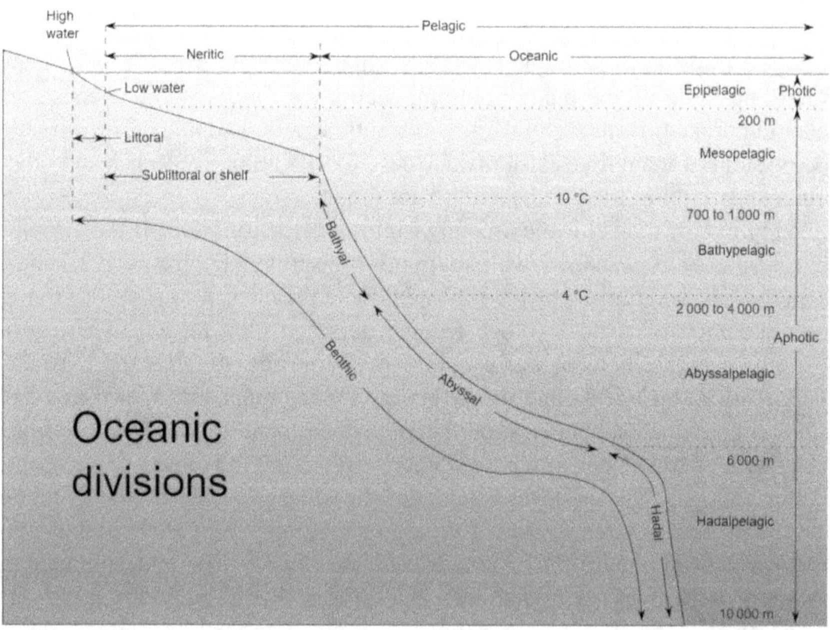

Figure 7.8 Oceanic Zones. See the text for descriptions. *Source*: Chris huh. Own work. Public domain.

Ocean-based economic opportunities for which markets don't exist yet, such as marine ecosystem services, are estimated to be worth at least $24 trillion.[7]

The oceanic zone is an extremely complex system, so it is useful to divide it into five "zones" for discussion. These zones overlap in many cases and may vary somewhat in different parts of the world. Figure 7.8 shows these zones.

The five zones of the open ocean are as follows:

1. The **Epipelagic Zone** or the "sunlight zone," is the area from the surface to 200m deep. Light can penetrate through this layer so photosynthesis can take place, which supports a thriving algae community. Algae that live in the epipelagic zone are responsible for much of the original food production for the entire ocean and create at least 50 percent of the oxygen in the atmosphere. The zone contains many microscopic plankton that form the basis of oceanic food webs. These plankton include *phytoplankton* (primary producers for most aquatic food webs), *zooplankton* (secondary consumers, single-celled to large invertebrates like fish larvae and jellyfish), and *ultraplankton* (photosynthetic bacteria). The epipelagic is home to many iconic animals, such as whales and dolphins, billfishes, tunas, jellyfishes, and sharks, many of which are of enormous economic importance.
2. The **Mesopelagic Zone** (also known as the "Twilight Zone") is the area between 200-1000m deep. Light is very limited in this zone, and there is less oxygen available to organisms. The mesopelagic zone occupies about 60 percent of the earth's surface and about 20 percent of the ocean's volume. It hosts a diverse biological community that includes ultraplankton and zooplankton, along with larger crustaceans, fish, squid, and many kinds of gelatinous animals.
3. The **Bathypelagic Zone** (also known as the "midnight zone") lies between 1000 and 4000m deep. Light cannot penetrate these depths leaving this region of the ocean in complete darkness. The zone is characterized by a steady temperature of around 4°C (39°F). Organisms in this zone must be able to function in cold temperatures and withstand extreme hydrostatic pressure, which ranges from 100 to 400 atmospheres (1500–6000 psi). Organisms include some fish, squids, ctenophores (comb jellies), and jellyfish. Many organisms use bioluminescence (light produced by specialized cells in the body) to attract food and mates.
4. The **Abyssopelagic Zone** extends from 4000 to 6000m deep and is commonly known as The Abyss due to the general absence of life found at these great depths. There is no light and little oxygen at these depths. Pressure can be as high as 750 atmospheres (11,000 psi). The major food source for organisms is dead organic matter from upper layers that slowly sinks through the water to form "marine snow." Organisms here

often have bizarre shapes and include fishes, squid, and ctenophores, many of which are bioluminescent.

5. The **Hadopelagic Zone** is characterized by deep sea canyons and trenches exceeding depths of 6000m. This includes the famous Mariana Trench, which contains the deepest known point on the earth's surface, the Challenger Deep, with a depth of 11,034 meters. The hadopelagic zone is characterized by extreme depth and pressure (up to 1100 atmospheres, over 16,000 psi), temperatures that hover just above freezing, little oxygen, and a complete absence of sunlight. Organisms include crustaceans and mollusks, and even a few fishes like eels, along with bacteria and some zooplankton. Most of these organisms are detritus feeders, which eat dead organic material from the bottom. Many organisms are bioluminescent. The hadopelagic zone is home to *hydrothermal vent* communities, which support a rich, unusual system discussed below.

Hydrothermal Vent Communities

Hydrothermal vents occur where hot, mineral-rich gases escape from deep in the earth through cracks in the seafloor. These gases mix with seawater, creating dark, hot clouds with concentrated nutrients. This chemical cloud gives the vents the popular name "black smokers" (figures 7.9a and 7.9b). Certain bacteria have the ability to utilize chemical reactions with these materials to form energy. This energy formation from chemicals, called *chemosynthesis*, is an alternative to photosynthesis, so these chemosynthetic bacteria are the producers in hydrothermal vent communities, taking the place of plants in traditional sunlight-based ecosystems. Bacteria, along with other organisms, use chemicals to erect elaborate structures.

Figure 7.9a Hydrothermal Vents the "Candelabra" Black Smoker at a Water Depth of 3,300 Meters in the Logatchev Hydrothermal Field on the Mid-Atlantic Ridge

Figure 7.9b A Living Community at a Hydrothermal Seep on the Mid-Ocean Ridge at a Water Depth of 3030 m. *Source*: By MARUM – Zentrum für Marine Umweltwissenschaften, Universität Bremen—Marum, CC BY 4.0

With chemosynthetic bacteria as producers, a wide range of organisms create the communities that we associate with hydrothermal vents. These communities include bacteria and other microscopic organisms, filter feeders like tube worms, and a variety of crustaceans, mollusks, and fishes.

7.4 FRESHWATER ECOSYSTEMS AND TRANSITION ZONES

Freshwater ecosystems represent only a small portion of the total water on the surface of the earth but have a disproportionate effect on life as we know it. Freshwater ecosystems cover less than 1 percent of earth's surface but are home to at least 10 percent of earth's species. More than 68 percent of the fresh water on earth is found in icecaps and glaciers, and just over 30 percent is found in groundwater. Only about 0.3 percent of our freshwater is found in the surface water of lakes, rivers, streams, and freshwater wetlands.

Freshwater ecosystems can be either *lentic* meaning that they have little or no regular flow, like lakes and ponds, or *lotic* meaning that they have regular flow, like rivers and streams.

Lakes, Ponds, Rivers, and Streams

These are some of the most recognizable freshwater ecosystems. They provide many services to humans, including drinking water, transportation, recreation, and a home to a wide variety of organisms.

There are differences, of course, in the characteristics of lentic systems like lakes and ponds and lotic systems like rivers and streams. We will examine lake and pond ecosystems first.

Lake and Pond Ecosystems

We should begin by saying that there is no clear agreement on the differences between a lake and a pond. Most ecologists agree that ponds are smaller than lakes (less than 5 hectares, or 12 acres, in surface area), have a maximum depth of 5 meters, and less than 30 percent emergent vegetation.[8] For our purposes, the differences are mostly just a difference in scale, so we will discuss them together as "lakes." That said, we must acknowledge that ponds represent biologically important parts of ecosystems.

The earth has a variety of freshwater lake types, from ponds to Lake Baikal in Siberia, which is almost a mile in depth and holds 23,000 cubic kilometers of water (equal to the total volume of the Great Lakes in the United States and Canada).

Most lakes contain fresh water, but some, especially those where water cannot escape via a river, can become saline (salt) lakes over time. Some lakes, such as the Great Salt Lake in Utah, are saltier than the oceans.

Most lakes support a variety of aquatic organisms, but some do not. For example, the Dead Sea in the Middle East is too salty for nearly all aquatic life.

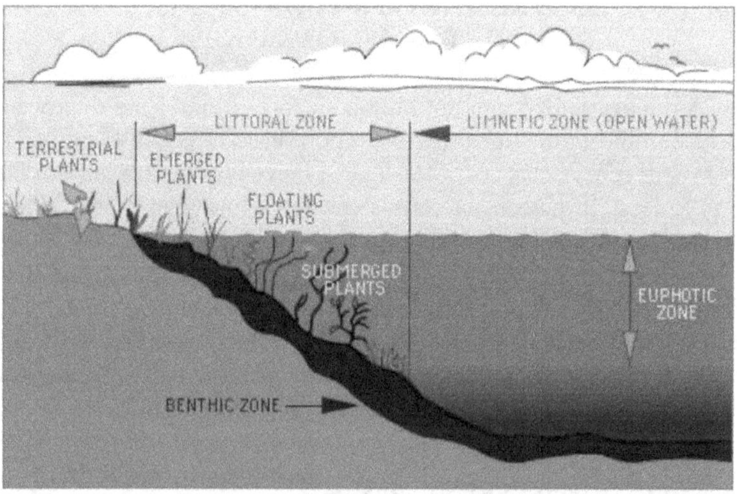

Figure 7.10 Zonation of a Pond or Lake Ecosystem. See the text for a description of the zones. *Source*: B. Bhakta, Michigan State University Extension—December 12, 2012. Inland lake habitats critical to maintaining healthy lake ecosystems. https://www.canr.msu.edu/news/inland_lake_habitats_critical_to_maintaining_healthy_lake_ecosystems

Lakes formed by erosion from glaciers, such as the Great Lakes, can be thousands of feet deep. Lakes also vary in depth. Some very large lakes may be relatively shallow, such as Lake Pontchartrain in Louisiana, which has a maximum depth of about 15 feet.

Like the open ocean, lakes have a characteristic zonation. A lake can have five distinct zones (figure 7.10).[9] These zones are based on the ability of light to penetrate:

- **Euphotic zone** refers to the area from the surface down to the depth where light becomes too low for organisms to undergo photosynthesis. The exact depth varies because it is dependent on water clarity. The euphotic zone has two subzones:
  - **Littoral zone** refers to the nearshore area that is the shallowest and warmest part of the lake where sunlight penetrates all the way to the bottom sediments and allows rooted aquatic plants (macrophytes) and attached algae to grow. This zone is most affected by changes at the shoreline.
  - **Limnetic zone** refers to the area of open water where the lake is too deep to allow rooted plant growth. This zone contains free-floating organisms such as phytoplankton and zooplankton. Stronger swimmers such as fish can utilize the limnetic zone as well as the littoral zone.
- **Profundal zone** refers to the area in a deep lake where no light reaches, so photosynthesis is not possible. Bacteria use oxygen to break down organic matter. Some profundal zones can be completely devoid of oxygen, which results in few to no animals using this area.
- **Benthic zone** refers to bottom sediment where a diverse variety of activities take place. It may be made up of sand, mud, marl, and/or a mixture of any or all of these materials. Organic sediments serve as a nursery, refuge, and foraging area for fish, amphibians, birds, aquatic insects, and crustaceans. The productivity of the benthic zone varies and depends on factors such as the organic content of the sediment, the physical topography, and, in some cases, the rate of fish predation.

Depending on factors such as depth, temperature, characteristics of the soils, and local climate, lake ecosystems can have vastly different organisms. Much like terrestrial ecosystems, lakes in tropical areas often have much higher biodiversity than those in temperate areas.

Lakes are also classified according to their energy levels and light penetration. The three categories are:

- **Oligotrophic lakes:** (figure 7.11) These are generally clear and cold. The lake substrate is typically firm and sandy. Nutrient levels are low, so the lake generally does not support large populations of aquatic plants, animals, or algae.

- **Mesotrophic lakes:** These have moderate amounts of nutrients and contain healthy, diverse populations of aquatic plants, algae, and fish.
- **Eutrophic lakes:** (figure 7.12) These are high in nutrients and contain large populations of aquatic plants, algae, and fish. The lake substrate is typically soft and muddy. The aquatic plants and algae often grow to nuisance levels, and the fish species are generally tolerant of warm temperatures and low-oxygen conditions.

Figure 7.11 An Oligotrophic Lake. Oligotrophic lakes are clear but nutrient poor. *Source*: Mirror Lake near Mt. Hood, Oregon. U.S. Forest Service. https://www.fs.usda.gov/recarea/mthood/recreation/recarea/?recid=53566

Figure 7.12 A Eutrophic Lake. This eutrophic lake in South Carolina is largely covered by tiny floating plants called duckweed. *Source*: Photo by the author

Some ecologists recognize a fourth category of *hyper-eutrophic lakes*. These are very high in nutrients and often exhibit large algae blooms, which may include dangerous levels of blue-green algae. Most hypereutrophic lakes are small impoundments of streams and are fed by large watersheds composed of urban and/or agricultural land uses. Notably, agricultural and other human practices often allow nutrients like fertilizers with high phosphorus levels to enter lakes and contribute to what is called *cultural eutrophication*.

The federal CWA §314 requires that most lakes in the United States be classified according to the Trophic State Index (TSI) system, which is based on the characteristics that we discussed. The EPA and other federal agencies use the formulation of the TSI in table 7.1.

River and Stream Ecosystems

Most Americans live within a mile of a river or a stream. Rivers and streams provide a variety of ecological, economic, and cultural values. River and stream ecosystems vary from high-elevation, stony streams, to flowing channels used by ships and boats, to shallow wetland deltas. They have striking regional differences that create distinct habitats. The bed of a river or stream can be stony or soft, lush with underwater vegetation, murky or clear. Each type of river or stream provides an environment for many different species of organisms, as well as different life stages.

River and stream ecosystems are the lotic (flowing) versions of lakes and ponds. Like lakes and ponds, they vary greatly in their characteristics depending on factors such as climate and substrate. But unlike lakes and ponds, rivers and streams have a flow rate which is determined in large part by local topography. Steeper inclines produce much higher flow rates. Most of us are familiar with the rapidly flowing "white water" often found in mountain streams (figure 7.13).

The terminology used for rivers and streams varies depending on the source. Technically, a stream is a continuous body of surface water that flows within the bed and banks of a channel. Depending on its location and

Table 7.1 Relationships between Trophic State Index, chlorophyll, phosphorus, Secchi depth, and trophic class

Trophic State Index	Chlorophyll (μg/L)	Phosphorus (μg/L)	Secchi depth (m)	Trophic Class
< 30—40	0—2.6	0—12	> 8—4	Oligotrophic or hypotrophic
40—50	2.6—7.3	12—24	4—2	Mesotrophic
50—70	7.3—56	24—96	2—0.5	Eutrophic
70—100+	56—155+	96—384+	0.5— < 0.25	Hypertrophic

Figure 7.13 Big South Fork of the Cumberland River in Kentucky. Angel Falls, a Class IV rapids. *Source*: An Chuck Summers. National Park Service.

characteristics, a stream may be referred to by a variety of local or regional names. Long, large streams are usually called rivers, while smaller, less voluminous, and more intermittent streams are known as streamlets, brooks, or creeks. For our purposes, we can think of rivers as large streams. The study of rivers, streams, and waterways in general is known as surface hydrology.[10]

Many streams in arid regions are dry during times of the year with less rainfall. Sudden storms can produce sudden and destructive flooding. However, any river or stream can be subject to flooding. The frequency and severity of flooding are increasing due to climate change, and we will examine these issues more thoroughly in chapter 12.

Rivers and streams are important as conduits of the water cycle, instruments in groundwater recharge, habitats for organisms, and corridors for fish and wildlife migration. The biological habitat in the immediate vicinity of a stream is called the *riparian zone*. Rivers and streams play an important corridor role in connecting fragmented habitats and thus in conserving biodiversity.[11]

While rivers and streams vary depending on local conditions, they have a characteristic zonation, as demonstrated in figure 7.14. From the ecosystem perspective, rivers and streams have three distinct habitat areas:

- **Riverbeds**, or the water channel itself.
- **Riverbank riparian zones**, which include the land, trees, and water-loving animals and plants along the channel.
- **Floodplains**, or the low, flat land spreading out from the channel. This area periodically floods during heavy rains and snowmelt. Floodplains may remain wet for long periods of time, creating rich wetland habitats.

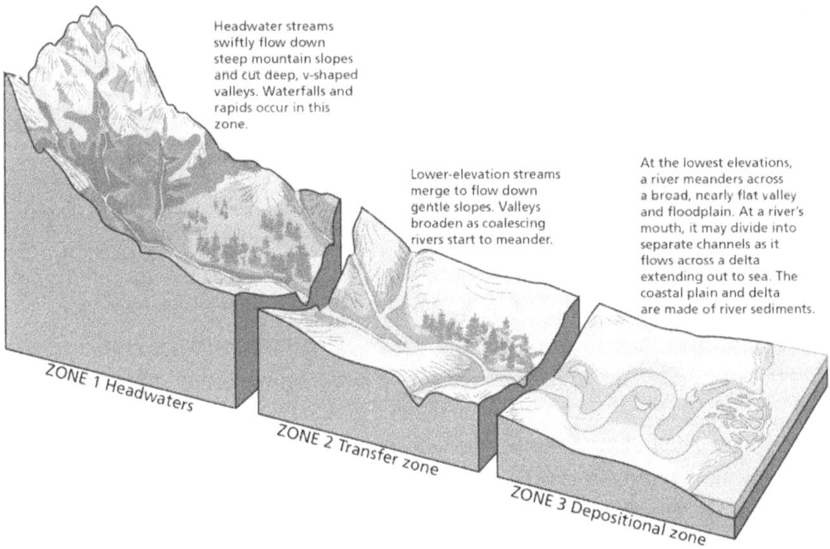

Figure 7.14 Zones of a River Ecosystem. *Source*: Trista L. Thornberry-Ehrlich, Colorado State University. National Park Service. https://www.nps.gov/subjects/geology/fluvial-landforms.htm

Rivers and streams provide important services for organisms. They are home to abundant fish and wildlife, including freshwater fishes, migratory fish like salmon and trout, invertebrates that provide food for fishes and other organisms like birds, and many endangered and threatened species.

Rivers provide important benefits to humans, too. These ecosystem services include irrigation for crops and clean water for consumption (60 percent of American drinking water comes from rivers). Hydroelectric power supplies much of the world with electricity. Historically, rivers have played an important role in transportation, bringing grain, coal, ore, and imports to market, and providing an important pathway for individual transportation. Rivers and streams are also important to recreation and tourism.

Rivers and streams are also important in flood protection. During periods of precipitation, floodplains can absorb large volumes of water. This provides natural flood control for coastal communities, preventing billions of dollars in damages.

Humans have had important historical impacts on rivers and streams. We have built dams for power and levees for shipping, dredged channels for navigation, and canals for irrigation. We've built towns and cities along banks and washed our wastes downstream.

Dams, culverts, and other barriers block migratory fishes from returning to their historic spawning grounds. When fish can't reach their habitat, they can't reproduce and maintain or grow their populations. Dams also alter water flow and sediment deposits, changing ecological conditions both above and below a dam. Structures like levees can control flooding in one area but increase flood risk in another. Disconnecting rivers from their floodplains increases the risk of extreme flooding and landslides.

Human activities can degrade or destroy important river and floodplain habitats. Here are some examples:

- Hardened shorelines increase water velocity, increasing erosion and negatively affecting the habitat for organisms.
- Digging channels or straightening rivers destroys nearby floodplains and wetlands, and can lead to flooding. Riverside development can reduce shade, which can lead to warmer waters that threaten many species.
- Paved surfaces also increase polluted runoff from roadways, parking lots, and roofs. Overused rivers can run dry long before they reach the sea, destroying important habitats.
- Polluted runoff into rivers and streams can cause serious damage to downstream ecosystems and destroy plant and animal communities. Important contributors are outdated farming methods that drain irrigation water containing fertilizers, herbicides, and pesticides into rivers and streams. Excess nutrients and toxic chemicals can then concentrate in waterways, causing algal blooms and dead zones where aquatic life cannot survive.
- Excess stormwater can cause sewage systems to overflow and allow untreated human waste into rivers. These overflows create disease risks and add nutrient pollution. The resulting algae overgrowth can be toxic to fish and people.

Global climate change has only made these threats more serious. We will discuss these issues in chapter 12.

Freshwater Wetlands

Freshwater wetlands are classified the same way as saltwater wetlands. In other words, a freshwater marsh is dominated by grasses (figure 7.15) , and a freshwater swamp is dominated by trees (figure 7.16). Other freshwater wetland types include bogs and fens with peat deposits, and prairie potholes that were formed by retreating glaciers in the upper Midwest.

Some freshwater wetlands cover many thousands of square kilometers, such as the Everglades in Florida, which is 11,100 square kilometers (4,300 square miles) in extent. The Everglades was larger in historic times before large-scale draining began in the nineteenth century, but it is dwarfed by the Pantanal (Brazil, Bolivia, and Paraguay) with a surface area that ranges, depending on the source, from 140,000 km^2 (54,000 mi^2) to 220,000 km^2 (85,000 mi^2). Other

Figure 7.15 A Freshwater Marsh Dominated by Nattail plants (Typha) near Indiana Dunes National Lakeshore. *Source*: Chris Light. Own work. Wikimedia Commons. https://creativecommons.org/licenses/by-sa/4.0/deed.en

Figure 7.16 A Bald Cypress-Tupelo Swamp in South Carolina. The projections from the water are cypress knees, which probably function as pneumatophores for aeration. *Source*: Photo by the author

notable freshwater wetlands are the Ngiri-Tumba-Maindombe (Democratic Republic of the Congo) at 66,000 km^2, the Okavango Delta (Botswana) at 55,300 km^2, and the Esteros del Iberá (Argentina) at 45,000 km^2.

Freshwater wetlands have high nutrient levels and are some of the most energetically productive ecosystems on earth. They can sustain a vast array of plant communities that, in turn, support a wide variety of wildlife. As a result, freshwater wetlands sustain a diversity of life that is disproportionate to their size.

In addition to their considerable value as habitat, freshwater wetlands serve to mitigate flood damage and filter excess nutrients from surface runoff. They are called "nature's kidneys"!

Unfortunately, freshwater wetlands have suffered major losses due to human actions. Some have been degraded by excessive deposits of nutrients and sediment from construction and farming. Many others have been intentionally filled, mostly for agriculture since the soils are extremely rich. Severe flooding and nutrient deposition to downstream waters often follow wetland destruction and degradation. This realization has spurred the protection and restoration of freshwater wetlands, as we will discuss below.

Transition Zones between Saltwater and Freshwater

Some of the most important aquatic ecosystems in the world are found where freshwater rivers flow into saltwater oceans (figure 7.17). These areas are known as estuaries, river deltas, and some saltwater wetlands. The water here tends to be brackish, or a mixture of freshwater and saltwater.

Many different habitat types occur in and around estuaries, including shallow open waters, freshwater and saltwater marshes, swamps, sandy beaches, mud and sand flats, rocky shores, oyster reefs, mangrove forests, river deltas, tidal pools, and seagrass beds.

Figure 7.17 The Snohomish River Estuary in Washington State. *Source*: NOAA Fisheries. https://www.fisheries.noaa.gov/feature-story/5-lovely-reasons-why-we-heart-estuaries

Estuaries are under tidal influence, so the depth of water fluctuates depending on the tide. They are protected from the full force of ocean waves, winds, and storms by landforms such as barrier islands or peninsulas.

Estuarine ecosystems are some of the most energetically productive on earth. They create more organic matter each year than almost any area of comparable size on earth. The sheltered waters of estuaries also support unique communities of plants and animals specially adapted for life at the margin of the sea. They are important areas for the migration and reproduction of many species of organisms.

Thousands of species of fish, birds, mammals, and other animals depend on estuarine habitats for shelter, food, and a place to reproduce. Most marine organisms, including most commercially important species of fish, depend on estuaries at some point during their development.

Because they are energetically productive, estuaries provide ideal areas for migratory birds to rest and refuel during their migration. Because many species of fish and wildlife rely on the sheltered waters of estuaries as protected spawning places, estuaries are often called the "nurseries of the sea."

Estuaries perform many other valuable services. Water draining from uplands carries sediments, nutrients, and pollutants to estuaries. As the water flows through wetlands such as swamps and salt marshes, much of the sediments and pollutants are removed. This filtration process creates cleaner and clearer water, which benefits both marine life and humans.

Salt marsh grasses and other estuarine plants also help prevent erosion and stabilize shorelines.

Wetland plants and soils also act as natural buffers between the land and ocean, absorbing floodwaters and dissipating storm surges. This protects upland habitat as well as valuable real estate from storm and flood damage.

Estuaries have important value to humans, too. Their resources provide economic benefits for tourism, fisheries, and recreational activities. The protected coastal waters of estuaries also support important public infrastructure, serving as harbors and ports vital for shipping and transportation.

The economy of many coastal areas is based primarily on the natural beauty and bounty of estuaries. When those natural resources are imperiled, so too are the livelihoods of those who live and work in estuarine watersheds. Over half the U.S. population lives in coastal areas, including along the shores of estuaries. Coastal watershed counties provided sixty-nine million jobs and contributed $7.9 trillion to the Gross Domestic Product in 2007.[12]

The same threats that we saw to rivers and streams above apply to estuaries, along with additional threats that we associate with the degradation of wetlands.

There are many reasons that aquatic ecosystems deserve protection. We will examine the legal mechanisms that protect aquatic ecosystems in chapter 8.

* * *

Review Questions

1. Explain why Earth can be described as a "water planet."
2. Contrast the composition and organisms that are found in a typical sandy versus rocky shoreline ecosystem.
3. Describe the organisms that make up a coral ecosystem, and where on the planet they are located.
4. Saltwater marshes and swamps are some of the energetically richest ecosystems in the world. Explain why.
5. List the five layers of open ocean ecosystems and describe the conditions and organisms of each layer.
6. Describe chemosynthesis and its significance to hydrothermal vents.
7. List the five layers of a freshwater lake and describe the conditions and the organisms of each layer.
8. Contrast the composition and organisms that are found in oligotrophic and eutrophic lakes.
9. Describe the characteristics and organisms in each of the three zones of a river ecosystem.
10. Contrast the composition and the organisms of a freshwater marsh and a freshwater swamp.
11. Explain the significance of aquatic transition zones like estuaries.

Case Analysis

Many residents and businesses around the United States rely on groundwater, accessed by wells, as their primary source of drinking water. When groundwater becomes contaminated with pollutants, it is often unusable as drinking water. Residents who are affected by groundwater contamination have several legal avenues against the polluters, including common-law actions, which we discussed in chapter 2.

The following is an excerpt from a New Jersey Supreme Court decision from 1995 that contains several common law causes of action.[13] The case was complicated and involved issues including insurance claims and vicarious liability, but we will limit our analysis to the common-law actions involved.

> From 1959 to 1975, Rule owned and operated a gas station at 930 Lacey Road in Lacey Township . . .
>
> In December 1984, residents of Barnegat Pines complained to the Ocean County Health Department (OCHD) about odors in their well water. OCHD tested a number of wells, finding fourteen to be contaminated with gasoline-related volatile organic compounds (VOCs). VOC concentrations found in many residential wells exceeded NJDEP's acceptable limits. The NJDEP

circumscribed a contaminated area covering fifteen blocks and determined that Rule's Service Station was the most likely source of contamination.

. . .

At trial, plaintiffs sought to prove that discharges of gasoline and other petroleum products from the gas station between 1959 and 1975, when Rule operated the Texaco station, contaminated their wells. In support of their allegations, plaintiffs sought to introduce testimony of Albert D. Young, a petroleum distribution consultant and retired Exxon executive.

. . .

At a "Rule 8" hearing, conducted under Evidence Rule 8 (now N.J.R.E. 104(a)), Young testified that he believed that above-ground gasoline spills "probably happened more frequently than not" during Rule's ownership of the station. According to Young, in the 1960s, as the capacity of fuel trucks increased, the trucks commonly pumped more fuel into the tanks than they could hold, thereby causing overflow and spills. He theorized that such spills, which were common during the 1960s and 1970s, had occurred at Rule's station. Young postulated that soil contamination in the areas around the underground tank and fuel islands indicated that spills had occurred during the delivery of gas or the pumping of fuel to customers' cars between 1959 and 1981. Young admitted on cross-examination, however, that he had never visited the service station, had never spoken with Rule or any of his employees, knew nothing about the operating practices at the station, and knew of no spills or leaks having occurred at the station from 1959 to 1981.

The trial court excluded Young's testimony as a "net opinion." Although the court found Young qualified as an expert on petroleum-product distribution, it concluded that he was not qualified as a hydrologist. Hence, Young was unqualified to render an opinion on the likelihood that any of the spills would reach the ground water. Because Young could not exclude spills that occurred in the period from August 1975, when the Ritchies purchased the station, to 1981, when they replaced the tanks, he could not render an opinion indicating Rule as the source of contamination.

. . .

In its charge, the court instructed the jury that it could find Rule liable under theories of negligence or strict liability. Further, the jury could find Texaco strictly liable, if it first found that Texaco had owned the underground tanks and pipelines and that they had leaked. Finally, the court instructed that the jury could find Texaco vicariously liable, if it found that Rule had been negligent and an apparent-agency relationship existed between Rule and Texaco.

. . .

Plaintiffs also argued that the trial court erroneously charged the jury that the Spill Act imposed strict liability for discharges from underground tanks only. According to plaintiffs, the Spill Act applies to discharges above, as well as those below, the ground. Although the Appellate Division agreed that the trial court had erred, it found the error harmless. The Appellate Division acknowledged that the Spill Act creates a private cause of action for recovery of damages caused by discharges of petroleum products, but held that the Act limits recovery to clean-up and removal costs. We agree. Such a cause of action does not extend, however, to damages arising from emotional distress, enhanced risk of disease, loss of enjoyment of property, and other economic and financial harm.

. . .

We affirm the Appellate Division substantially for the reasons stated in its opinion.

Questions

1. What common-law actions are involved in this case? Can you think of other common law causes of action that were not included?
2. It's hard to tell from the excerpt, but the plaintiff property owners lost in the trial court, the appellate court, and the California Supreme Court. On what legal basis did they lose?
3. Poor Mr. Young, the property owners' expert witness, was largely excluded from consideration in the case. What was the court's reason for excluding most of his testimony? Was the court correct in doing that?
4. California has a law called the Spill Act[14] that allows state authorities to direct spill responses and cleanup, and to authorize natural resource damage assessment and restoration. The plaintiffs in this case attempted to use that law as a means to recover damages. Did it work? Why, or why not?
5. Did the California Supreme Court get this right? Justify your conclusion.

NOTES

1. Several notable recent publications have explored the significance of Earth as a water planet. For example, see: Gaskin-Reyes, C. (Ed.). 2016. *Water Planet: The Culture, Politics, Economics, and Sustainability of Water*. ABC-CLIO.

2. Archived as UN Atlas of the Oceans. 2010. https://web.archive.org/web/20131105071822/http://www.oceansatlas.org/servlet/CDSServlet?status=ND0xODc3JjY9ZW4mMzM9KiYzNz1rb3M~.

3. See UN Environment Programme, ISU, ICRI and Trucost. 2018. *The Coral Reef Economy: The Business Case for Investment in the Protection, Preservation and Enhancement of Coral Reef Health*, p. 36. https://www.unep.org/resources/report/coral-reef-economy.

4. USEPA. *Definition of a Wetland.* https://www.epa.gov/wetlands/what-wetland.

5. See NOAA National Ocean Service. *What is a Salt Marsh?* https://oceanservice.noaa.gov/facts/saltmarsh.html.

6. See: McKee, K.L. 2004. *Global Change Impacts on Mangrove Ecosystems.* USGS Publications Warehouse. https://pubs.usgs.gov/publication/fs20043125.

7. See the United Nations Conference on Trade and Development (UNCTAD). *2022 Report "5 Global Actions Needed to Build a Sustainable Ocean Economy."* https://unctad.org/news/5-global-actions-needed-build-sustainable-ocean-economy.

8. Holgerson, M.A., Richardson, D.C., Roith, J., Bortolotti, L.E., Finlay, K., Hornbach, D.J., et al. 2022. Classifying Mixing Regimes in Ponds and Shallow Lakes. *Water Resources Research, 58*, e2022WR032522. https://doi.org/10.1029/2022WR032522.

9. This discussion is based on: Bhakta, B. 2012. *Inland Lake Habitats Critical to Maintaining Healthy Lake Ecosystems.* Michigan State University Extension, December 12. https://www.canr.msu.edu/news/inland_lake_habitats_critical_to_maintaining_healthy_lake_ecosystems.

10. See Langbein, W.B., & Iseri, K.T. 1995. *Hydrologic Definitions: Stream. Manual of Hydrology: Part 1. General Surface-Water Techniques (Water Supply Paper 1541-A).* USGS.

11. There are many excellent sources of information on stream and river hydrology, for example: Alexander, L.C., Autrey, B., DeMeester, J., Fritz, K.M., Golden, H.E., Goodrich, D.C., . . . McManus, M.G. 2015. *Connectivity of Streams and Wetlands to Downstream Waters: Review and Synthesis of the Scientific Evidence* (Vol. 475). EPA/600/R-14. https://cfpub.epa.gov/ncea/risk/recordisplay.cfm?deid=296414.

12. See Kildow, J. T., Colgan, C.S., & Scorse, J.D. 2009. *State of the US Ocean and Coastal Economies 2009.* National Ocean Economics Program. Publications 4. https://cbe.miis.edu/noep_publications/4

13. *Bahrle v. Exxon Corp.*, 652 A.2d 178 (N.J. Super. Ct. App. Div. 1995), aff'd 678 A.2d 178 (N.J. Sup. Ct. 1995).

14. Actually, the Lempert-Keene-Seastrand Oil Spill Prevention and Response Act of 1990, CA Govt Code §8670.1 (2023).

Chapter 8

Protecting Aquatic Ecosystems

Chapter Learning Outcomes

At the end of this chapter, the reader will be able to:

1. *Describe the historical origins of the federal Clean Water Act.*
2. *Distinguish between point sources and nonpoint sources of water pollution.*
3. *Describe the National Pollutant Discharge Elimination System (NPDES) permit system.*
4. *Explain new source performance standards (NSPS) in the Clean Water Act.*
5. *Discuss water quality standards under the Clean Water Act.*
6. *Explain how the Safe Drinking Water Act protects underground drinking water supplies.*
7. *Describe laws that protect the marine environment.*
8. *Discuss federal laws that protect marine fisheries.*
9. *Explain the dangers of plastic pollution and possible solutions.*

We learned a lot about the ecology of aquatic ecosystems in chapter 7, including the values that they have for us and our planet. They are some of the most important ecosystems on the earth, but they are also among the most vulnerable. Let's turn our attention to the ways that the legal system can protect aquatic ecosystems, or at least limit the damage.

The logical place to begin our inquiry is with the federal Clean Water Act (CWA), the primary federal law that regulates water pollution, especially in freshwater ecosystems. Its stated objective is "to restore and maintain the chemical, physical, and biological integrity of the Nation's waters."

Figure 8.1 Firefighters Battle a Fire on Ohio's Cuyahoga River in 1952. NOAA. National Ocean Service. *Source*: Cleveland Press Collection at Cleveland State University Library. https://oceanservice.noaa.gov/facts/earth-day.html

This statement is a recognition by Congress of many decades of abuse of U.S. waterways that culminated in tragedies like the fires of oil slicks on several U.S. rivers. For example, the heavily polluted Cuyahoga River in Ohio caught fire on several occasions between 1936 and 1969, when debris and oil had concentrated on the water's surface and ignited (figure 8.1). A blaze in 1969 came at a time of increasing environmental awareness and symbolized years of environmental neglect.

8.1 THE FEDERAL CLEAN WATER ACT

The Clean Water Act of 1977 (CWA) has a long and interesting history. Before we begin our analysis of the relevant parts of the law, let's explore that history.

Clean Water Act History and Policy Issues

The earliest contribution to what would become the CWA, and arguably the first environmental law in the United States, was the 1899 Rivers and Harbors Act.[1] Also known as the *Refuse Act*, the Rivers and Harbors Act was passed at a time when the major corridors for commerce in the United States were the nation's waterways. Prior to the late 1800s, these waterways were also a convenient place for the dumping of trash and debris with virtually no controls. In the 1880s, Congress recognized that these waterways had become fouled over time, and that interfered with commerce.

The most significant parts of the Rivers and Harbors Act for our purposes are Sections 9 and 10, which require authorization from the Secretary of the Army, acting through the Army Corps of Engineers, for the deposit of any obstructions or the construction of any structure in or over any navigable water of the United States.[2] The significance of the Refuse Act is that it established the authority of the federal government to regulate the disposal of refuse into the navigable waters of the United States. As we will see, this authority was used for the enforcement of what came to be known as "effluent limitations" in the modern Clean Water Act.

The second contributor to the modern CWA is the Federal Water Pollution Control Act of 1948,[3] which was the first major U.S. law to address water pollution. The Act authorized the Surgeon General of the Public Health Service, in cooperation with other Federal, state, and local entities, to prepare comprehensive programs for eliminating or reducing the pollution of interstate waters and tributaries and improving the sanitary condition of surface and underground waters. It also authorized the Federal Works Administrator to assist states, municipalities, and interstate agencies in constructing treatment plants to prevent discharges of inadequately treated sewage and other wastes into interstate waters or tributaries.[4] We can think of the 1948 Act as establishing a system to regulate how polluted a body of water can be. We will see the influence of the 1948 Act in establishing "water quality standards" in the modern CWA.

In the end, Congress opted for the "effluent limitations" approach, although water quality standards are an important backup. Enforcement is by the National Pollutant Discharge Elimination System (NPDES) permit program, which uses technology forcing standards to make reasonable further progress in eliminating pollution of the nation's waters. We will explore these standards below.

Growing public awareness and concern for controlling water pollution in the 1960s led to sweeping amendments in 1972. As amended in 1972, the law became commonly known as the Clean Water Act (CWA). Since 1972, the original statute has been amended extensively through the years.

Several amendments to the CWA bear further scrutiny. The amendments of 1977 introduced three new categories of pollutants: toxic, conventional, and nonconventional pollutants. These designations allow different regulations for water pollutants that are considered to be more dangerous to people and ecosystems. A summary of these useful categories is in table 8.1.

Subsequent amendments in 1981 streamlined the municipal construction grants process, improving the capabilities of treatment plants built under the program. Additional amendments continue to change details of the CWA to authorize additional water quality programs, modify standards and procedures to govern allowable discharges, and provide funding for construction grants or general program funding.

A significant amendment to the CWA was the Oil Pollution Act of 1990, which provides the basic statutory authority for oil and hazardous substances

Table 8.1 Categories of Pollutants Introduced in CWA Amendments of 1977

1. Toxic pollutants: Based on consideration of toxicity, persistence, degradability, and effects on organisms.
2. Conventional pollutants: Pollutants traditionally regulated as discharges from POTWs, such as biological oxygen demand (BOD), suspended solids, fecal coliform bacteria, pH, oil, and grease.
3. Nonconventional pollutants: Pollutants that are neither toxic nor conventional, such as ammonia, chlorides, nitrates, and color.

pollution prevention, contingency planning, and response activities within the 200-mile Exclusive Economic Zone.

What Waters Are Covered by the CWA?

An important initial observation is that the CWA does not apply to all water in the United States. Under a Supreme Court rule issued in 2023, all waters (such as streams, oceans, rivers, and lakes) with "a continuous surface connection" to "navigable waters" are covered under the CWA.[5] This rule significantly reduced the coverage of the CWA by excluding waters that are connected by intermittent seasonal water bodies or waters that are connected by subsurface (groundwater) connections. As we will see, the effect is to exclude many of the nation's wetlands and intermittent streams from CWA protection.

The 1972 statute frequently uses the term "navigable waters" but also defines the term as "waters of the United States, including the territorial seas."[6] Regulations interpreting the 1972 law have included water features such as intermittent streams, prairie potholes, sloughs, and wetlands as "waters of the United States," but these definitions are in a state of legal limbo.

Other statutes regulate other sources of water in the United States. For example, the Safe Drinking Water Act, discussed below, regulates underground aquifers and other drinking water supplies.

Structure of the Clean Water Act

The Clean Water Act is a complex statute that requires legally trained professionals to interpret, but the basic structure is relatively simple. A good place to start is to understand the difference between point sources and nonpoint sources of water pollution, because the CWA treats them differently than nonpoint sources.

Let's start with *point sources* of water pollution. CWA Section §502(14) defines a point source as "any discernible, confined, and discrete conveyance, including but not limited to any pipe, ditch, channel, tunnel, conduit, well . . . or vessel or other floating craft, from which pollutants are or may be discharged."[7] It's obvious that what Congress had in mind is something like what we see in figure 8.2. Pipes like these were a familiar sight in pre-CWA America.

There are numerous examples of point sources of water pollution. Factories, including oil refineries, pulp and paper mills, and chemical, electronics,

Figure 8.2 Discharge from a Paper Mill in Jay, Maine in 1973. *Source*: USEPA. https://catalog.archives.gov/id/550713

and automobile manufacturers, typically discharge one or more pollutants (called *effluents*) in their discharged waters. In addition, discharges from wastewater treatment facilities, discharges from concentrated animal feeding operations (where animals are confined and fed), contamination from leaking septic systems, chemical and oil spills, and illegal dumping are all examples of point sources of water pollution.

Interestingly, agricultural stormwater discharges and return flows from irrigated agriculture are specifically exempted from the definition of point sources.[8] This was likely a political decision made in order to attract the necessary support from Congress for the Clean Water Act in 1972.

8.2 THE NPDES PERMIT SYSTEM

CWA Section 402 prohibits any person from discharging pollutants from a point source into the waters of the United States without a National Pollutant Discharge Elimination System (NPDES) permit. The NPDES permits contain limits on what can be discharged, monitoring and reporting requirements, and other provisions to ensure that the discharge does not decrease water quality or health.

CWA Section 301 prohibits the discharge of any pollutant to waters of the United States except in compliance with certain sections of the Act, including CWA Section 402. Section 402 established the NPDES permit program to be administered by the EPA or authorized states, territories, or eligible tribes.

The NPDES permit program provides two types of permits, individual and general, which are used to authorize point source discharges of pollutants to waters of the United States. Individual permits are issued by individual states or the EPA to a single facility and require submission of a permit application. General permits are developed by individual states or the EPA to cover classes or categories of dischargers under a single permit. General permits usually require facilities seeking permits to submit a notice of intent (NOI) to be covered, the contents of which are described in the general permit. Both types of permits are issued for a fixed period of time not to exceed five years.[9]

NPDES permits are issued by the Water Quality Protection Division of the EPA. The Water Enforcement Branch ensures that all discharges comply with NPDES permits. Many states have assumed responsibility for directly permitting and enforcing the NPDES program, but the EPA still oversees state-run NPDES programs.[10]

The CWA is called a *technology forcing* statute because of the demands placed on those who are regulated by it to achieve higher and higher levels of pollution abatement. New point sources and existing point sources of pollutants have different NPDES regulations. New sources are subject to more rigorous effluent limits than existing sources based on the concept that it is cheaper to minimize effluent pollutants if environmental controls are considered during plant design and construction than if an existing facility is retrofitted, which can be time-consuming and expensive. New source discharges must comply with standards based on the performance of demonstrated technology with the greatest degree of effluent reduction. These *new source performance standards (NSPS)* should represent the most stringent numerical values attainable. NSPS are based upon the best available demonstrated control technology and are at least as stringent as the best available technology.

When the original CWA was passed in 1972, industries were given until 1977 to install *"best practicable control technology" (BPT)* to clean up waste discharges. Municipal wastewater treatment plants were required to meet an equivalent goal, termed "secondary treatment," by that date. The focus of BPT technology was on controlling discharges of conventional pollutants, such as suspended solids, biochemical oxygen-demanding materials, fecal coliform and bacteria, and pH. These pollutants are substances that are biodegradable (i.e., bacteria can break them down), occur naturally in the aquatic environment, and deplete the dissolved oxygen concentration in water, which is necessary for fish and other aquatic life.

The CWA mandated greater pollutant cleanup than BPT by no later than 1989, generally requiring that industry use the *"best available technology" (BAT)* that is economically achievable. BAT-level controls generally focus on toxic substances and represent a stricter, higher level of technology than BPT. Compliance extensions of as long as two years are available for industrial sources utilizing innovative or alternative technology. Failure to meet statutory deadlines could lead to enforcement actions.

A separate type of permit is required to dispose of dredged or fill material in the nation's waters, including wetlands. The CWA Section 404 permit program is administered by the U.S. Army Corps of Engineers, subject to and using EPA's environmental guidance. Some types of activities are exempt from permit requirements, including certain farming, ranching, and forestry practices which do not alter the use or character of the land; some construction and maintenance; and activities already regulated by states under other provisions of the act. EPA may delegate certain Section 404 permitting responsibility to qualified states and has done so twice (Michigan and New Jersey). For some time, the wetlands permit program has been one of the most controversial parts of the law. Some who wish to undertake development projects in wetlands maintain that federal regulation intrudes on and impedes private land-use decisions, while environmentalists seek more protection for remaining wetlands and limits on activities that are authorized to take place in wetlands.

Nonpoint Source Regulations

Nonpoint sources of pollution, which the EPA, many states, and most environmental scientists believe are responsible for the majority of water quality impairments in the nation, are not subject to CWA permits or other regulatory requirements under federal law (figure 8.3). They are covered by state programs for the management of runoff, under Section 319 of the act.[11]

Congress chose not to address nonpoint sources through a regulatory approach, unlike its actions with "point" sources. Instead, when it added Section 319 to the CWA in 1987, it created a federal grant program that provides money to states, tribes, and territories for developing and implementing NPS management programs.

Under CWA Section 319, states, territories, and delegated tribes are required to develop nonpoint source pollution management programs in order to receive Section 319 funds. Once it has approved a state's nonpoint source program, EPA provides grants to these entities to implement NPS management programs under Section 319(h). Section 319 is a significant source of funding for implementing NPS management programs, but there are other federal (e.g., Farm Bill), state, local, and private programs.

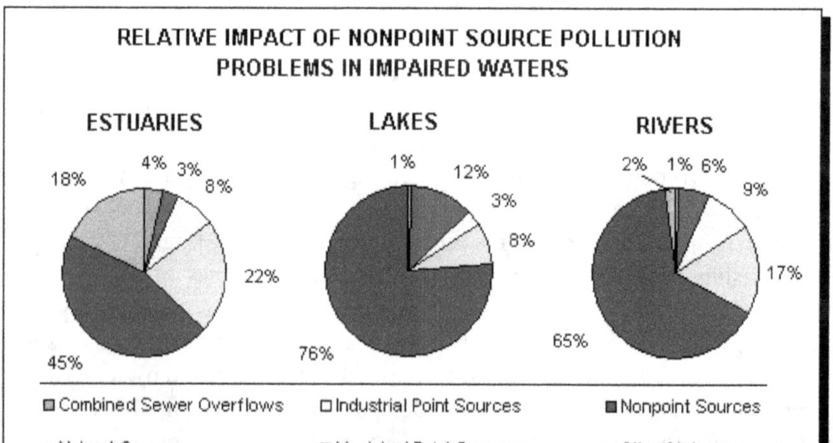

Figure 8.3 Nonpoint Source Pollutants, such as Sediments, Nutrients, Pesticides, Fertilizers, and Animal Wastes, Account for more than Half of the Pollution in U.S. Waters. *Source*: David J. Welsch, for the United States Department of Agriculture. https://commons.wikimedia.org/wiki/File:Percentages_sources_of_pollution_that_impair_water_bodies

Water Quality Standards

Water quality standards are scientifically measurable standards that quantify the goals of the CWA and include designated uses, water quality criteria, and antidegradation policies. Designated uses are the uses of a waterbody that society would like to achieve in a location, such as fishing and swimming, industrial water supply, or agricultural irrigation. Water quality criteria are the conditions necessary to achieve a waterbody's designated uses. Antidegradation policies are rules implemented by states to prevent waterbodies with higher-than-normal water quality from becoming impaired in the future.

Several policies must be followed when assigning designated uses to water bodies. Any use that the water body has achieved since November 28, 1975, even if it is not possible at present, must be included in the designated uses for that water body. Because the broad goals of the CWA include fishing and swimming, every water body must include those designated uses unless the downgrading process has been followed to allow the reduction of quality in a water body. Designated uses may only be reduced by following a downgrade process that demonstrates that the water quality goals are unattainable because of natural background conditions, irreversible human-caused conditions resulting in substantial environmental harm, or excessive social and economic cost. Under EPA regulations, an existing use may never be removed from the designated uses for a water body.

Water quality criteria are technical conditions that are required to reach the designated uses in a particular water body, including concentrations of pollutants, temperature, pH, and "no toxic chemicals in toxic amounts." However, acceptable water quality criteria can also include narrative descriptions that are difficult to quantify scientifically. These include no objectionable bottom deposits, no unsightly floating debris, and no nuisance odor or color. The CWA requires that water quality criteria be established based on the latest scientific knowledge.

Section 303(d) of the CWA authorizes the EPA to assist states, territories, and authorized tribes in listing impaired waters and developing Total Maximum Daily Loads (TMDLs) for these waterbodies. A TMDL establishes the maximum amount of a pollutant allowed in a waterbody and serves as the starting point or planning tool for restoring water quality. A TMDL is the calculation of the maximum amount of a pollutant allowed to enter a waterbody so that the waterbody will meet and continue to meet water quality standards for that particular pollutant. A TMDL determines a pollutant reduction target and allocates load reductions necessary to the source(s) of the pollutant.

An antidegradation policy is a required component of CWA water quality standards that states are required to adopt and enforce. These policies are designed to protect water bodies with water quality that is higher than standards require and prevent degradation or loss of existing uses. EPA regulations direct states to include three levels of protection in their antidegradation policies. The first level protects existing uses and requires that the level of water quality necessary to sustain those uses be maintained. The second level requires that states maintain water quality that exceeds levels necessary to support recreation and wildlife uses unless a lower water quality is necessary to allow important economic or social development. Finally, the third level of antidegradation policy requires that states maintain water quality in high-quality waters that are considered "outstanding National resource[s]," including waters within national and state parks, wildlife refuges, and waters with "exceptional recreational or ecological significance."[12]

Does the Clean Water Act Work?

There is strong disagreement as to whether the CWA has been effective in achieving its goals, and whether the costs to industry and development have been worth the benefits in cleaner water.

In general, it is clear that the CWA has been successful at reducing pollution that enters our rivers and lakes from point sources. However, nonpoint source pollution is still a significant problem for clean water. Polluted runoff is a significant source of pollution for many of our rivers, lakes, and streams across the country.

8.3 THE SAFE DRINKING WATER ACT

The Safe Drinking Water Act (SDWA) is the primary federal law intended to ensure safe drinking water for the public.[13] Under the SDWA, the Environmental Protection Agency is required to set standards for drinking water quality, and oversee all states, localities, and water suppliers that implement the standards.

Also administered under the SDWA is EPA's Underground Injection Control (UIC) Program, which regulates the construction, operation, permitting, and closure of injection wells that place fluids underground. Among the UIC Program's objectives are to:

- Protect public health through the protection of underground sources of drinking water (USDWs).
- Provide a means for industries, municipalities, and small businesses to dispose of wastes and store fluids.

Public Drinking Water Systems

The SDWA applies to every public water system (PWS) in the United States. There are currently over 148,000 public water systems providing water to almost all Americans at some time in their lives (figure 8.4). The SDWA does not cover private wells (in 2020, 13 percent of U.S. households were served by private wells).[14] The SDWA also does not apply to bottled water, which is regulated by the Food and Drug Administration (FDA), under the Federal Food, Drug, and Cosmetic Act.

The SDWA's primary regulatory mechanisms are the National Primary Drinking Water Regulations, which identify maximum contaminant levels (MCLs) or treatment techniques generally on a contaminant-by-contaminant basis. MCLs and treatment techniques are generally based on the potential for toxicity to humans and the feasibility of detection and water treatment. Not all potential contaminants are regulated under this program, however. SDWA requires an ongoing program for evaluation and regulation of new contaminants found in drinking water.

Under the SDWA, EPA establishes minimum standards for tap water supplied by public water systems. A public water system is defined as a water system that has fifteen service connections or regularly serves twenty-five individuals for at least 60 days per year. Public water systems obtain their

Figure 8.4 A Public Drinking Water Supply. *Source*: Wikimedia Commons. https://commons.wikimedia.org/wiki/File:Drinking_water_fountain.jpg

water from two sources, groundwater and surface water. Depending on the quality of the source water, pre-treatment may be required prior to distribution to the public for consumption.

National Primary Drinking Water Regulations (NPDWR) are standards and treatment techniques that public water systems must follow. These regulations protect public health by limiting contaminant levels in drinking water. National Secondary Drinking Water Regulations (NSDWR) are guidelines to help public water systems manage their drinking water for issues not related to health, such as taste, color, and smell. Water systems are not required to follow these water quality standards for the fifteen contaminants listed. Although these contaminants may not be harmful to public health, if they are in water at levels above the standards, they can cause the water to look cloudy or colored, or to taste or smell bad.

The SDWA includes a process that EPA must follow to name unregulated contaminants that may require regulation in the future. EPA must publish this list of contaminants—called the "Contaminant Candidate List," or CCL—every five years and decide whether to regulate at least five or more of the contaminants on the list (called "Regulatory Determinations").

The SDWA's National Primary Drinking Water Regulations may apply when hydraulic fracturing or other oil or natural gas production activities impact a public water system's supply (e.g., during release cleanup or pre-treatment of groundwater or surface water sources contaminated by exploration or production activities). However, these regulations do not apply to the oil or natural gas extraction and production activities themselves.

Underground Injection

The SDWA requires the EPA to regulate the construction, operation, permitting, and closure of injection wells that place specific fluids underground through the Underground Injection Control (UIC) Program (figure 8.5). Oil and natural gas operators use injection wells for three purposes:

- Disposal of fluids brought to the surface in connection with natural gas storage operations or oil and gas production activities (including produced water, hydraulic fracturing fluids, etc.)
- Enhanced recovery of oil or natural gas, for example waterflood operations and hydraulic fracturing.
- Injection of hydrocarbons for storage provided the hydrocarbons are liquid at standard temperature and pressure. The injection of natural gas for storage, if the gas is of pipeline quality, is exempt from regulation under the SDWA.

There are six categories or "classes" of injection wells distinguished by the fluids injected, activities, construction, injection depth, design, and operating techniques.[15]

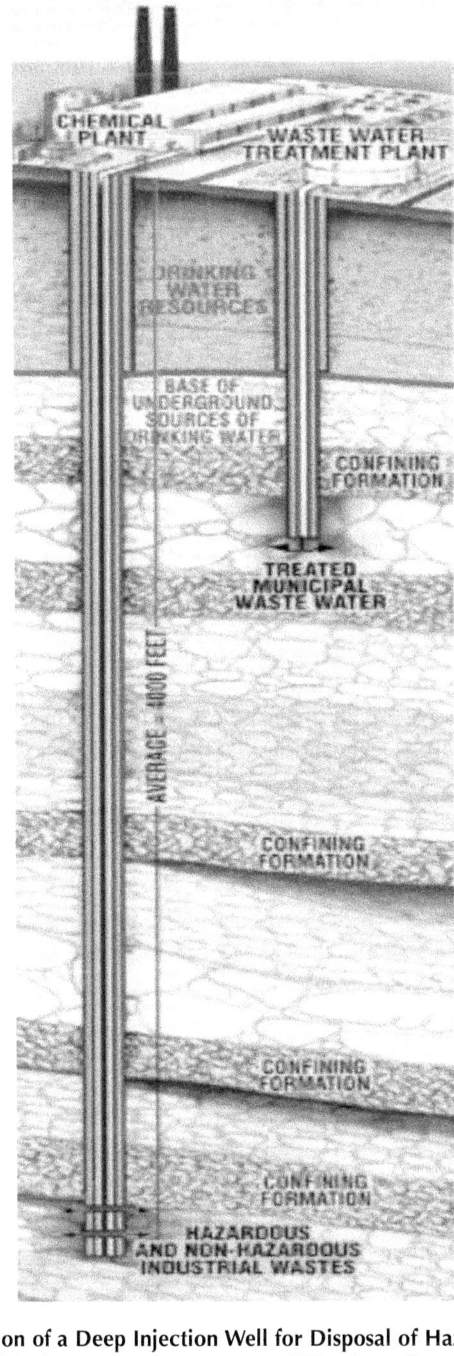

Figure 8.5 Illustration of a Deep Injection Well for Disposal of Hazardous, industrialIn-dustrial, and Municipal Wastewater. *Source*: USEPA. https://commons.wikimedia.org/wiki/File:Deep_injection_well.jpg

The SDWA specifically excludes hydraulic fracturing from UIC regulation under SDWA Section 1421(d)(1); however, the underground injection of fluid that includes diesel fuel during hydraulic fracturing is still regulated by the UIC Program. Diesel fuels may contain a number of chemicals of concern, including benzene, toluene, ethylbenzene, and xylene compounds (BTEX), which are highly mobile in groundwater and are regulated because of the risks they pose to human health. For a discussion on what constitutes diesel fuels, please see the EPA Guidance on Diesel Fuels Hydraulic Fracturing.

There are many other environmental laws that impact water quality. For example, hazardous chemicals are a frequent source of water pollution, and they are regulated by other federal statutes, such as the Resource Conservation and Recovery Act (RCRA). We will discuss hazardous chemicals and the RCRA in later chapters.

8.4 REGULATION OF MARINE ECOSYSTEMS

We saw in chapter 7 that marine, saltwater ecosystems are unique in many ways, including environmental characteristics and organisms. It should come as no surprise that marine ecosystems require unique protection from environmental threats. We will explore several of the most important of these regulatory mechanisms.

As a starting point, recall that the major environmental threats to the world's oceans are from water pollution, marine habitat loss and degradation, overfishing, and the effects of climate change. Let's explore how some of our major environmental laws and regulations address these threats.

Before we examine the marine applications of several important environmental laws, let's first consider which oceans are under U.S. jurisdiction. Most of the open oceans of the world are under international jurisdiction, and their protection is mostly a matter of various international treaties and agreements. This level of protection is difficult to enforce, and the protection of international waters is often inadequate.

The extent to which U.S. environmental laws extend into the oceans is complex and is largely beyond the scope of this book. However, as a general rule, U.S. laws apply to internal waters (adjacent to the coasts), to territorial seas (12 nautical miles), and probably to the contiguous zone (to 24 nautical miles). These ocean zones are shown in figure 8.6.

In 1982, the United Nations Convention on the Law of the Sea[16] developed the concept of an exclusive economic zone (EEZ), which provides countries with sovereign rights for the purpose of exploring, exploiting, conserving, and managing natural resources of the seabed, subsoil, and waters above it. The United States claimed a 200-nautical-mile EEZ in 1983 (figure 8.7). While federal environmental laws do not apply directly to EEZs, they provide important opportunities for the protection of marine ecosystems.

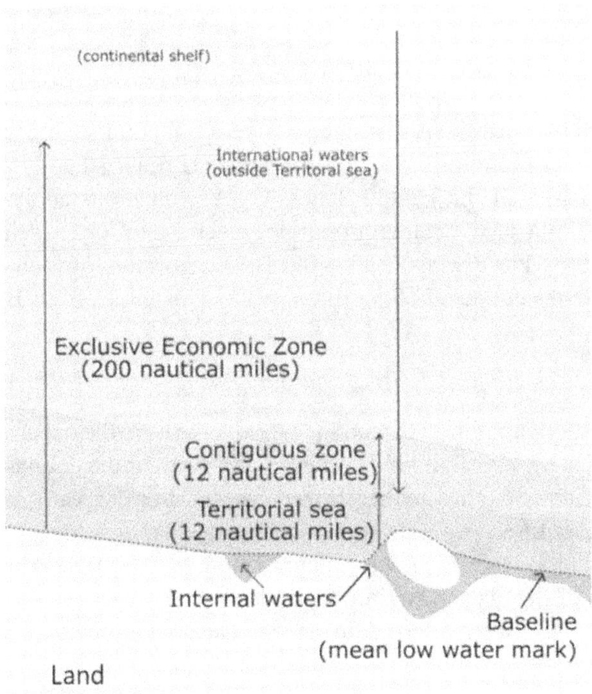

Figure 8.6 Currently Recognized Zones of the Ocean. *Source*: Wikimedia Commons. https://creativecommons.org/licenses/by-sa/3.0/deed.en

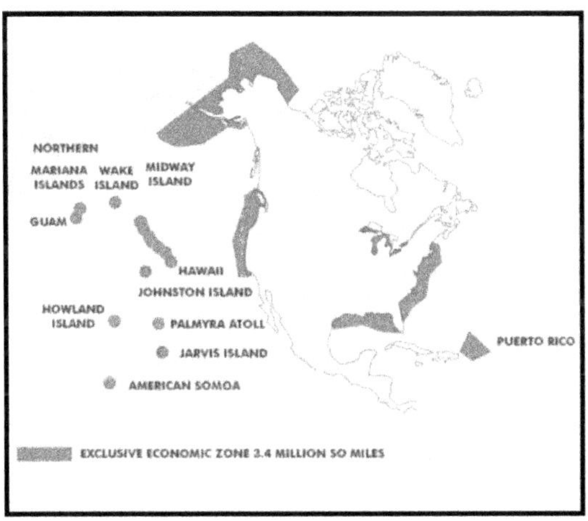

Figure 8.7 The Exclusive Economic Zone of the United States. NOAA. National Ocean Service. *Source*: Public domain. https://oceanservice.noaa.gov/facts/eez.html

Laws that Prevent Ocean Pollution

The Clean Water Act

Recall that the federal CWA distinguishes between point source and nonpoint sources of water pollution. Since oceans under U.S. jurisdiction are included among the waters of the United States, CWA protections apply. The provisions of the CWA have been an important mechanism for charging and punishing polluters. For example, after the BP Deepwater Horizon oil disaster, the United States collected $5.5 billion in CWA penalties from BP.[17]

The Marine Protection, Research, and Sanctuaries Act (MPRSA)

For centuries, communities around the world have used the oceans as a dumping ground for wastes of all kinds. Marine dumping, or the deliberate disposal at sea of wastes or other matter from vessels, aircraft, platforms, or other man-made structures, often destroys or degrades habitats for aquatic species and causes coastal and open ocean degradation, which affects the health and productivity of the marine environment.

The MPRSA,[18] also referred to as the Ocean Dumping Act, generally prohibits:

- Transportation of material from the United States for the purpose of ocean dumping.
- Transportation of material from anywhere for the purpose of ocean dumping by U.S. agencies or U.S.-flagged vessels.
- Dumping of material transported from outside the United States into the U.S. territorial sea.

Vessels must obtain a permit issued by the USEPA to deviate from these prohibitions.

Under the MPRSA, the standard for permit issuance is whether the dumping will "unreasonably degrade or endanger" human health, welfare, or the marine environment. The EPA is charged with developing ocean dumping criteria to be used in evaluating permit applications.

The MPRSA implements the requirements of the London Convention,[19] which is the international treaty governing ocean dumping. Among other things, the London Convention regulates the dumping of polluted bilge water from ships, which is a major source of coastal pollution, especially near ports (figure 8.8). Discharging oily bilge water may also run afoul of the Clean Water Act's prohibition against discharging oil into the waters of the United States.

Figure 8.8 Bilge Water Is Discharged from a Ship. U.S. Coast Guard. *Source*: Public domain image. https://commons.wikimedia.org/wiki/File:Ship_pumping_ballast_water.jpg

In addition to the CWA and the MPRSA, there are other federal laws that address ocean pollution, either directly or indirectly. For example, the Oil Pollution Act of 1990, which is actually a set of amendments to the Clean Water Act, strengthened the EPA's ability to prevent and respond to catastrophic oil spills.[20] A trust fund financed by a tax on oil is available to clean up spills when the responsible party is incapable or unwilling to do so.

Another law that helps to prevent marine pollution is the Coastal Zone Management Act of 1972, which encourages collaboration between state governments and the federal government "to preserve, protect, develop, and where possible, to restore or enhance, the resources of the Nation's coastal zone for this and succeeding generations."[21] The CZMA provides federal funding for states with an approved Coastal Zone Management Plan to act as stewards of the coastal zone with the power to halt federal actions inconsistent with their Plan. These Plans create an added level of protection for coastal regions and give states control in maintaining coastal integrity, especially in regulating nonpoint source pollution that may find its way into oceans.

Laws that Protect Species and Habitats

We discussed the federal Endangered Species Act[22] and the Marine Mammal Protection Act[23] at length in chapter 4. These laws have provided crucial protections to vulnerable marine species within U.S. jurisdiction for nearly a half-century. Recall that both laws contain prohibitions against the "taking" of protected species, and the protection of "critical habitat" that may include marine areas.

Other important federal laws also serve to protect marine species and habitats, including those outside the continental United States. The National Marine Sanctuaries Act (NMSA)[24] of 1972 was enacted with the purpose of designating important marine environments as sanctuaries for their environmental, historical, or cultural significance. Currently, there are thirteen marine sanctuaries in the National Marine Sanctuary System. A Marine National Monument was designated by a presidential proclamation in 2006 as a conservation area. The Papahānaumokuākea Marine National Monument encompasses nearly 140,000 square miles of ocean waters and coral reefs around the northwestern Hawaiian Islands in the Pacific Ocean.

More recently, the Coral Reef Conservation Act (CRCA)[25] of 2000 was established to preserve coral reef ecosystems. The CRCA promotes responsible management to gain reliable information on the current condition of coral reefs. The CRCA established four major programs: National Coral Reef Action Strategy, Coral Reef Conservation Program, Coral Reef Conservation Fund, and the National Program. The National Coral Reef Action Strategy was developed to establish goals for research, monitoring, and conservation, along with addressing regional and international issues. The National Program assesses the conservation of coral reefs in various ways, such as monitoring and restoration. In addition, it enhances public awareness of coral reefs through educational programs. The National Program also facilitates cooperative work between federal, state, and regional efforts that are working to improve coral reef ecosystems.

Another recent development is the creation of Marine Protected Areas (MPAs) around the world, regions designated and managed for the long-term conservation of marine resources, ecosystem services, or cultural heritage. While the specifics differ around the globe, the United States defines a marine protected area as "any area of the marine environment that has been reserved by federal, state, tribal, territorial, or local laws or regulations to provide lasting protection for part or all of the natural and cultural resources therein."[26]

There are nearly 1,000 MPAs within United States coastal, oceanic, and Great Lakes waters, including the National Marine Sanctuaries and marine national monuments, as well as many national parks, national wildlife refuges, state parks, locally protected areas, and more.[27] While the effectiveness of MPAs is still under evaluation, there is hope in the scientific community that MPAs will be a valuable source of information and protection for vulnerable aquatic ecosystems.[28]

Laws that Protect Marine Fisheries

It is difficult to overstate the importance of the world's fisheries to human welfare. Fisheries and aquaculture provide food for billions of people and play an important role in the local economy of coastal communities in many countries (figure 8.9). Estimates suggest that about 58.5 million people are employed worldwide in primary fish production, of which approximately 21 percent are women.[29] Including subsistence and secondary-sector workers and their dependents, about 600 million livelihoods depend at least partially on fisheries and aquaculture. In 2019, aquatic foods provided about 3.3 billion people with at least 20 percent of their average intake of animal protein, with an even higher proportion in many developing countries.

In the United States, fisheries are an $853 million annual industry that accounts for 1.7 million jobs (figure 8.10).

Figure 8.9 A Fishing Boat with Nets Extended. *Source*: https://www.fisheries.noaa.gov/feature-story/united-states-promotes-labor-standards-improved-monitoring-south-pacific-regional

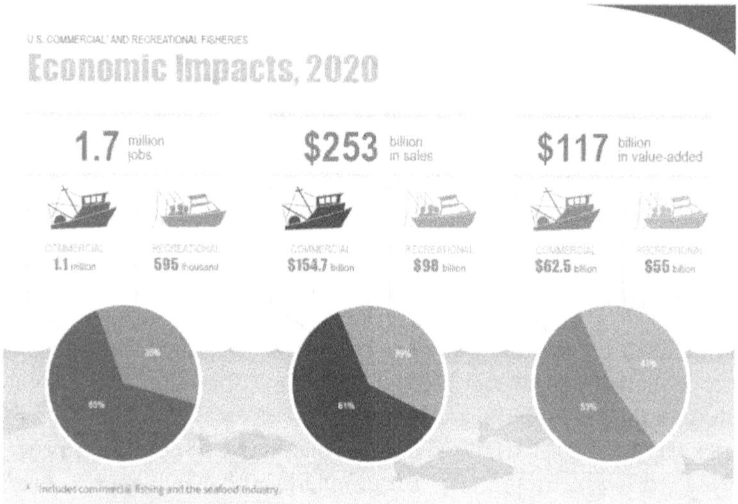

Figure 8.10 Commercial and Recreational Fisheries Impacts in 2020. *Source*: NOAA Fisheries. https://www.fisheries.noaa.gov/national/sustainable-fisheries/fisheries-economics-united-states

However, while ocean resources boost growth and wealth, they have been seriously impacted by anthropogenic factors. The United Nations estimates that, worldwide, the percentage of fishery stocks that are not within biologically sustainable levels rose from 10 percent in 1974 to 35.4 percent in 2019.[30] Marine and aquatic ecosystems are under stress from climate change, overfishing, and other unsustainable fishing practices, as well as from pollution from other human activities, as we have discussed.

Healthy fish stocks are a key component of biodiversity and ecosystems, and as such, are key to the ocean's capacity to provide essential ecosystem services.

In the United States, the regulation of commercial fisheries is regulated primarily by the Magnuson–Stevens Fishery Conservation and Management Act (MSA)[31] of 1976. It was originally adopted to meet the following goals:

- to create a U.S. fishery conservation zone extending out to 200 nautical miles off the U.S. coast;
- to phase out foreign fishing activities within this zone;
- to prevent overfishing, especially by foreign fleets;
- to allow overfished stocks to recover; and
- to conserve and manage fishery resources.

The MSA has been amended several times. The 1996 Sustainable Fisheries Act amendments[32] most notably added measures to protect essential fish habitat. The Magnuson–Stevens Fishery Conservation and Management

Reauthorization Act of 2006 (MSRA) amended the MSA to mandate the use of annual catch limits and accountability measures to end overfishing, provide for fishery management by a limited access program, and call for increased international cooperation.

Laws Related to Climate Change

We spoke briefly of climate change in chapter 6, and we will discuss it further in later chapters. However, the oceans of the world are under particular pressures from climate change, and it is relevant to discuss a few laws in response.

A significant driver of climate change is the increase of the greenhouse gas carbon dioxide (CO_2) from anthropogenic sources. In addition to the effects of CO_2 as a greenhouse gas, it has an additional effect on oceans. When CO_2

Figure 8.11 Some Marine Organisms that Require Calcium Carbonate to Build Shells or Other Structures. *Source*: NOAA, Ocean Explorer. https://oceanexplorer.noaa.gov/explorations

comes in contact with seawater, it forms carbonic acid (H_2CO_3), a weak acid that nevertheless causes acidification (a decrease in pH) in aquatic ecosystems. Ocean acidification has a variety of effects, but an important consequence is a decrease in available carbonate, which affects marine organisms that use carbonate to build shells and other structures, like shellfish, coral, and calcifying plankton (figure 8.11).

The development of U.S. law related to ocean acidification has proceeded slowly. The Federal Ocean Acidification Research and Monitoring Act (FOARAM) of 2009 provides for ocean acidification research and monitoring.[33] In accordance with the mandates of the FOARAM act, the National Oceanic and Atmospheric Administration (NOAA) has developed a program to deploy a series of buoys in the open ocean to monitor pH levels (figure 8.12).

At the international level, the Intergovernmental Panel on Climate Change (IPCC) was established by the United Nations and the World Meteorological Organization to provide the world with a leading scientific group and perspective on the current state and possible impacts of climate change, including changes to the oceans.[34]

Other effects of climate change are related to sea level rise. As global temperatures increase and ocean waters warm, the resulting loss of land-based ice (e.g., glaciers) causes sea levels to rise across the earth. There is concern that decreases in ocean salinity can affect ocean currents and habitats for marine

Figure 8.12 A NOAA Buoy, Deployed North of Iceland. It is equipped with a MAPCO2 monitoring system that measures CO_2 concentrations in the surface water and atmosphere every 90 minutes. *Source*: NOAA Research. https://research.noaa.gov/2013/09/05/a-more-acidic-arctic-noaa-deploys-first-buoy-in-region-to-monitor-levels-of-co2-absorbed-by-ocean/

organisms. The rise in sea levels has already caused increased coastal erosion, measurable loss of coastal wetlands, and habitat loss for marine species.[35]

8.5 THE SPECIAL PROBLEMS OF PLASTICS POLLUTION

Few environmental issues have drawn public interest as much as the issue of plastics pollution. While plastics pollution is a problem in all aquatic ecosystems, the threat to the oceans of the world is especially acute.

The scope of plastics pollution is huge. Plastics are ubiquitous in modern societies and are especially troublesome when used for packaging and other disposable uses. This plastic becomes an environmental pollutant when it is not properly recycled, incinerated, or kept in sealed landfills. Improperly treated plastic waste is truly a global problem affecting terrestrial and aquatic ecosystems worldwide (figure 8.13). One to two million metric tons of plastic waste enters the world's oceans yearly, affecting wildlife and ecosystems.

Plastics pollution affects aquatic ecosystems in three primary ways:

- Plastic waste, especially containers, may trap organisms or injure them if they are consumed (figure 8.14);
- Most plastic wastes take many decades to decompose but gradually break down into smaller and smaller fragments; and
- When plastics degrade, they frequently release harmful chemicals into the environment, such as PFAS and PFOA, the "forever chemicals" that can harm wildlife and ecosystems.[36]

Figure 8.13 Marine Debris Circulates through the Ocean and Can Pile Up on Coasts. *Source*: NOAA photograph. National Park Service, Oceans, Coasts & Seashores. https://www.nps.gov/subjects/oceans/ocean-plastics.htm

Figure 8.14 Stomach Contents of a Dead Albatross Chick. Includes a variety of plastic marine debris. Chris Jordan (via USFWS / CC BY 2.0). *Source*: Photo taken by Chris Jordan, uploaded by Foerster, Public Domain, https://commons.wikimedia.org/w/index.php?curid=26762401

Plastic pollution is usually classified in several ways. Based on size, it is classified as *macroplastics* (≥25mm), *mesoplastics* (<25–5mm), *microplastics* (<5–1mm), and *mini-microplastics* (<1mm–1µm). The smaller the size of the plastic particles, the more dangerous they are to wildlife.

Plastic pollution may also be classified as primary or secondary. Primary microplastics are those that enter the marine environment in their "micro" size. Secondary microplastics are those resulting from the breakdown of larger plastics in the marine environment. Microplastics are also classified into several shape categories, commonly named fragments, fibers, films, foam, and beads.

Unfortunately, most worldwide plastic pollution is unregulated. In the United States, over 80 percent of plastic waste ends up in landfills and is regulated by laws that we will discuss elsewhere, such as the Solid Waste Disposal Act. Laws that deal with water pollution in general usually cover plastic pollution to some degree. A proposed law called the "Break Free from Plastic Pollution Act" faces an uncertain future in Congress at this writing.

As we have seen, the threats to aquatic ecosystems are very real, and the solutions are not always obvious. But it is equally true that as a nation, the United States has made huge strides in aquatic ecosystem protection. We can do much better, of course, and future generations will have to struggle to continue our trajectory of improvement in the future.

* * *

Review Questions

1. Describe the roles of the Rivers and Harbors Act and the Federal Water Pollution Control Act in the structure of the Clean Water Act.
2. Distinguish between point sources and nonpoint sources of water pollution.
3. Explain the NPDES permit system of the Clean Water Act.
4. What are new source performance standards (NSPS), and how do they relate to point source regulation?
5. Describe water quality standards under the clean water act.
6. How does the Safe Drinking Water Act regulate public water supplies?
7. How does the Ocean Dumping Act regulate marine water pollution?
8. What federal laws protect marine fisheries?
9. What are the long-term consequences of climate change on marine ecosystems?
10. Describe the effects of plastics pollution on marine ecosystems.

Case Analysis

We discussed the Magnuson–Stevens Fishery Conservation and Management Act (MSA) above. Among other things, the MSA authorizes the National Marine Fisheries Service (NMFS) to set limits on the permissible catch of certain marine species. After a series of lawsuits in the mid-1990s, the NMFS entered into a settlement agreement with fisheries groups and set a quota for the collection of large coastal shark (LCS) stocks in the Atlantic Ocean and the Gulf of Mexico. Several environmental and conservation groups sued, arguing that the settlement agreement was illegal because the NMFS did not base its agreement on the best available science. The suit was filed in the Federal District Court for the Middle District of Florida. The following is an extract from the court's decision.[37]

> National Standard 2 requires that "[c]onservation management measures shall be based on the best scientific information available." 16 U.S.C. § 1851(a)(2). Under the national standard guidelines, "[s]cientific information includes, but is not limited to, information of a biological, ecological, economic, or social nature." 50 C.F.R. § 600.315(b) (1). "FMPs must take into account the best scientific information available at the time of preparation." 50 C.F.R. § 600.315(b) (2). When NMFS' decisions are reviewed for compliance with NS 2, the agency "warrants cautious deference in matters falling within [its] studied specialty and concerning which equivocal evidence and genuine scientific debate abound."
>
> Plaintiffs argue that NMFS' actions violated NS 2 in two distinct ways. First, Plaintiffs contend that NMFS' repeated protests that its December 2001 rule

was based on the best available science are patently false. Specifically, Plaintiffs argue that NMFS predetermined the outcome of the December 2001 rule by agreeing to the terms of the Amended Settlement Agreement, i.e., if the peer reviewers accepted the 1998 SEW Report, the 1999 quotas would take effect; if not, the 1997 quotas would remain in place. Plaintiffs allege that NMFS left itself no authority whatsoever to review the peer reviews and evaluate for itself what the best available scientific information might be. Rather, Plaintiffs contend, NMFS abdicated its legal duty to analyze and apply scientific information in implementing conservation management measures for LCS. Second, Plaintiffs argue that NMFS failed to comply with NS 2 because its decisions were not based on all available, current scientific information.

As for Plaintiffs' first argument that NMFS predetermined the outcome of the December 2001 rule and gave up its authority to evaluate for itself what constituted the best available scientific information, this Court finds that Federal Defendants did not violate NS 2.

Plaintiffs next argue that NS 2 was violated because NMFS' decisions were not based on all available, current scientific information. Plaintiffs contend that NMFS was required to consider information available through the time of its decisions in June and December 2001, but did not do so.

While NMFS did take the position in prior litigation that the scientific methods . . . used in the 1998 stock assessment that formed the basis of the management measures for the 1999 HMS FMP were in accordance with NS 2, subsequent analysis of those scientific methods caused NMFS to conclude that the projections of the 1998 SEW models no longer represent the best available science. In issuing the 2002 notification, NMFS explained:

[B]ased on the results of the independent peer reviews, NMFS no longer believes that the projections of the 1998 SEW models constitute the best available science. As a result, some of the management measures adopted in the HMS FMP that were based on the 1998 LCS SEW models are not appropriate at this time. Once a new LCS stock assessment is conducted and peer reviewed, NMFS will proceed with proposed and final rules to ensure any additional appropriate management measures based on this new information are in place to conserve the LCS stocks and its fishery.

Federal Defendants concede that they had scientific documents available to them prior to December 2001, but argue that they did not at that time have the opportunity to convene the SEW to evaluate and synthesize such information and make management recommendations thereon. NMFS chose to wait to consider new, long term management measures during the 2002 SEW. In addition, Federal Defendants point to the fact that the Plaintiffs did not identify any particular piece of scientific information cited in the 2002 Stock Assessment and Fishery Evaluation for Atlantic Highly Migratory Species that would undermine its decision to retain the 1997 quotas pending the reconvening of the SEW.

The Court finds that the decision of a majority of the independent scientific reviewers forms a rational basis for NMFS to conclude that the projections of the 1998 SEW models upon which the 1999 HMS FMP was based no longer represent the best available science. There is an obvious lack of consensus by and between the independent reviewers and the 1998 SEW scientists regarding the adequacy of the data and analyses in the 1998 SEW Report. This gives additional support to NMFS' decision to reject that data. Accordingly, the Court finds that the record contains a rational basis for maintaining the 1997 LCS quotas pending completion of a new SEW, and that Defendants did not violate NS 2 and are entitled to summary judgment on this claim.

Questions

1. The plaintiffs (conservation groups) argued that the NMFS did not use the best available science. On what do they base this argument?
2. There is a lot of discussion of "data." Where do these data come from?
3. The NMFS changed its mind and ended up rejecting the results of the 1998 SEW study. How did they justify the change?
4. In the end, the Court rejected the plaintiffs' arguments and supported the quotas established by the NMFS. Did the Court get it right? Can you think of reasons why the 1997 quotas should be changed?

NOTES

1. Actually, the Rivers and Harbors Appropriations Act of 1899, 33 U.S.C. § 401 et seq.
2. 33 U.S.C. §§ 401, 403.
3. 33 U.S. Chapter 26–Water Pollution Prevention and Control.
4. A summary of the 1948 Act is at the FedCenter site: https://www.fedcenter.gov/Bookmarks/index.cfm?id=2431.
5. Totenberg, N. 2023. The Supreme Court Has Narrowed the Scope of the Clean Water Act. *NPR*, May 25. https://www.npr.org/2023/05/25/1178150234/supreme-court-epa-clean-water-act
6. CWA §502 (7); 33 U.S.C. §1362.
7. 33 U.S.C. § 1362(14).
8. CWA § 502(14) states: "This term [point source] does not include agricultural stormwater discharges and return flows from irrigated agriculture."
9. Found in CWA § 402(b)(1)(B) and in the regulations at 40 CFR § 122.46.
10. CWA § 402(b)(1)(B) and 40 CFR 122.46.
11. CWA 319, 33 USC § 1329.
12. See EPA "Water Quality Standards." https://www.epa.gov/wqs-tech/what-are-water-quality-standards-0.
13. 42 U.S.C. § 300f et seq.
14. See EPA Private Drinking Water Wells. https://www.epa.gov/privatewells.
15. UIC regulations are at 40 CFR 144–148.

16. The United Nations Convention on the Law of the Sea (UNCLOS), adopted in 1982, produced a comprehensive set of international laws for the world's oceans and seas, establishing rules governing all uses of the oceans and their resources.

17. The USEPA's timeline of events related to the Deepwater Horizon oil disaster is at: https://www.epa.gov/enforcement/deepwater-horizon-bp-gulf-mexico-oil-spill.

18. 16 USC §1431 et seq. and 33 USC §1401 et seq.

19. 1972 Prevention of Marine Pollution by Dumping of Wastes and Other Matter (London Convention).

20. 33 U.S.C. §2701 et seq. (1990).

21. 16 U.S.C. §1451 et seq. (1972).

22. 16 U.S.C. §1531 et seq.

23. 16 U.S.C §§1361–1407.

24. 16 U.S.C. §1431 et seq.

25. 16 U.S.C. §6401 et seq.

26. Executive Order 13158 (5/26/2000).

27. A useful source of information on MPAs is NOAA's Ocean Exploration website at: https://oceanexplorer.noaa.gov/facts/mpas.html.

28. See Grorud-Colvert, K., et al. 2021. The MPA Guide: A Framework to Achieve Global Goals for the Ocean. *Science, 373*(6560), eabf0861. https://www.science.org/doi/10.1126/science.abf0861.

29. See the World Bank. *Blue Economy*. https://www.worldbank.org/en/topic/oceans-fisheries-and-coastal-economies.

30. Food and Agriculture Organization of the United Nations. *Towards Blur Transformation*. https://www.fao.org/state-of-fisheries-aquaculture.

31. 16 U.S.C. ch. 38 § 1801 et seq.

32. 16 U.S.C. §§ 1801–1884.

33. 33 USC §3701.

34. IPCC. https://www.ipcc.ch/.

35. For example, see the climate.gov website at https://www.climate.gov/news-features/understanding-climate/climate-change-global-sea-level.

36. Wee, S.Y., & Aris, A.Z. 2023. Revisiting the "Forever Chemicals," PFOA and PFOS Exposure in Drinking Water. *NPJ Clean Water, 6* Article 57. https://www.nature.com/articles/s41545-023-00274-6.

37. *Ocean Conservancy v. Evans*, 260 F. Supp. 2d 1162 (M.D. Fla. 2003).

Chapter 9

Solid and Hazardous Wastes

Chapter Learning Outcomes

At the end of this chapter, the reader will be able to:

- *List the types of solid and hazardous wastes.*
- *Describe the origins and disposal of household solid waste.*
- *Identify the special problems with electronic wastes.*
- *Discuss the modern concept of solid waste management.*
- *Describe the RCRA "cradle to grave" regulatory system for solid and hazardous wastes.*
- *Discuss the CERCLA remediation system for hazardous wastes.*
- *Identify the problems associated with brownfields, and possible solutions.*

The final step in our investigation of the "media" of the environment (air, water, and land) is an investigation of land. We learned a great deal about terrestrial ecosystems in chapters 3 and 5, so we can jump immediately into the consideration of the loss of terrestrial ecosystem functions due to contamination and habitat loss.

Sadly, severe contamination of terrestrial ecosystems by solid waste pollution is not an uncommon sight in many parts of the world. As seen in figure 9.1, virtually all ecosystem values can be destroyed, often irreparably.

We will begin by exploring sources of contamination of land by solid wastes and move to a discussion of the laws that regulate solid waste disposal and examine the mechanisms for remediating past contamination.

Figure 9.1 As Seen here, Virtually All Ecosystem Values Are Destroyed by Solid Waste Pollution. Shown here is the Mbale dumping site in the Eastern Region of Uganda. *Source*: Lubega Ibrahim, Own Work. Wikimedia Commons. https://creativecommons.org/licenses/by-sa/4.0/deed.en

9.1 WHAT ARE SOLID WASTES?

Before we begin our discussion of the effects of various wastes on ecosystems, we must first determine exactly what we mean by "solid waste." The World Health Organization (WHO) defines solid waste this way:

> *Solid waste is any type of garbage, trash, refuse or discarded material. It can be categorized according to where the waste is generated, for example as municipal solid waste, health care waste and e-waste. Over 2 billion tons of municipal solid waste are produced annually.*[1]

There are many ways to classify solid waste, of course.[2] We will learn later in this chapter that the regulatory classification of solid wastes is quite complex. A better place to start is with a version of the WHO classification of solid waste based on how they are generated. A useful list of categories of solid waste looks like this: municipal solid waste, industrial and commercial waste, agricultural waste, medical waste, e-waste, and hazardous waste. There is certainly overlap between these categories, but we will examine each.

- **Municipal solid waste** (MSW): This is household waste typically treated as garbage. We will discuss MSW more fully in section 9.2. MSW wastes are usually regulated by state laws and local ordinances.
- **Industrial and commercial waste:** Industrial waste results from industrial processes and manufacturing and consists of waste materials like used chemicals, solvents, scrap metal, and other byproducts from production

facilities. Commercial waste is produced by offices, schools, restaurants, shops, and other institutions, and includes waste such as paper, food waste, packaging, and other discarded items. Also included are construction and demolition debris, which includes concrete, wood, bricks, metals, and other materials used in building structures.

- **Agricultural waste:** These are generated by farming activities and include crop waste, animal manure, pesticides, and other waste produced in agricultural processes. Agricultural waste also includes wastes from livestock and poultry houses, and slaughterhouses.
- **Medical waste:** Wastes generated by healthcare activities, including used needles and syringes, pharmaceutical drugs, medical devices, soiled dressings, body parts, diagnostic samples, blood, chemicals, and radioactive materials. Medical waste is regulated mostly by state environmental and health departments, although the federal Occupational Safety and Health Act (OSHA) regulates all aspects of medical waste management that involve the safety of workers. OSHA's bloodborne pathogens standard has provisions to protect employees during containment, storage, and transportation of regulated medical waste.[3] Certain medical wastes are classified as **biohazards** if they might contribute to the spread of infectious diseases. Biohazards are collected in specially labeled bags and disposed of in approved state facilities.
- **E-waste:** These are discarded electronic devices and equipment, such as computers, televisions, smartphones, and appliances. E-waste can contain hazardous materials like lead, cadmium, beryllium, mercury, and other components that require specialized handling and disposal. The rapid worldwide increase in e-waste is due to frequent new model releases and unnecessary purchases of electrical and electronic equipment, short innovation cycles, low recycling rates, and a drop in the average life span of computers and other electronic devices.
- **Hazardous waste**. Materials that pose a significant risk to human health or the environment due to their toxic, flammable, corrosive, or reactive properties are considered hazardous. This category includes items like batteries, fluorescent bulbs, pesticides, certain chemicals, and medical waste. Any of the preceding categories of solid waste can contain hazardous waste. As we will discuss below, hazardous wastes are regulated primarily by the Resource Conservation and Recovery Act (RCRA).

Municipal Solid Waste (MSW)

MSW, commonly known as "trash" or "garbage," is the most recognizable form of solid waste to most residents of urban or suburban settings. MSW includes items discarded by single or multiple residential dwellings, hotels, schools, hospitals, businesses, and other similar permanent or temporary housing establishments or facilities. It consists of items that are used and then discarded, such as paper, plastics, food scraps, textiles, packaging materials, yard waste, and some household hazardous waste like batteries or cleaning products (figure 9.2). Plastic waste is an important part of solid waste, as we discussed in chapter 8.

Historically, most MSW was discarded in open dumps, in waterways, or was burned in open pits, all of which led to ecosystem loss and pollution of land, air, and water. Today, most MSW in the United States is discarded in *sanitary landfills*. These sanitary landfills are constructed with clay and plastic liners to prevent pollutants from leaking into surrounding soils and groundwater, and a leachate collection system that collects liquid leachate for treatment and disposal. They often include a methane gas recovery system that collects methane, which can be used to generate heat and electricity (figure 9.3).

Unfortunately, open dump sites remain a problem in many parts of the world (figure 9.1), with potentially tragic consequences for human health and the environment.

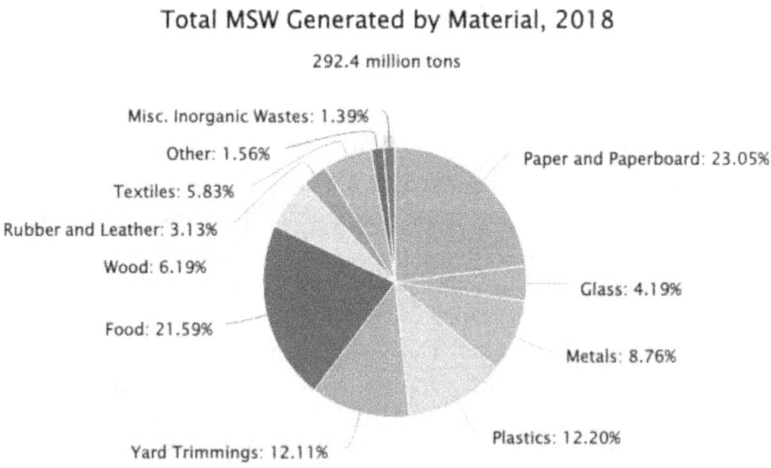

Figure 9.2 U.S. Department of Energy, National Energy Technology Laboratory. *Source*: 1.3.1 Waste Streams. https://netl.doe.gov/research/coal/energy-systems/gasification/gasifipedia/waste

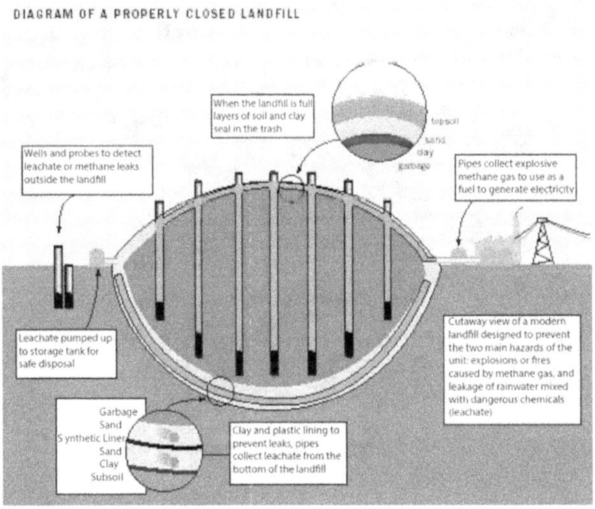

Figure 9.3 Structure of a Sanitary Landfill. U.S. EPA. What is a Municipal Solid Waste Landfill? *Source*: https://www.epa.gov/landfills/municipal-solid-waste-landfills

The Special Problems of E-wastes

Electronic wastes, or e-wastes, are discarded electronic devices and equipment, such as computers, televisions, smartphones, and appliances (figure 9.4). Americans own nearly three billion electronic products, or an average of twenty-four electronic products per household. With a constant supply of newer, faster electronic products on the market, older models are quickly replaced. As a result, electronics have become one of the fastest-growing segments of the solid waste stream.

The international scope of the e-waste problem is huge. A 2019 United Nations-backed report found that a record 53.6 million metric tons of electronic waste was discarded worldwide. That number is expected to increase to 74.7 million tons by 2030. This amount of e-waste generated could fill more than 100 Empire State buildings. The report also found that in 2019 only 17.4 percent of that e-waste was collected and recycled, which means 82.6 percent of e-waste was not formally collected or managed in an environmentally friendly manner.[4]

Electronic waste has become an issue of national and international importance due to the quantities produced, but also because of the various toxic materials and heavy metals it contains. Electronics often contain arsenic, chromium, cadmium, lead, mercury, beryllium, zinc, copper, silver, and gold. They also contain brominates as flame retardants. If crushed or improperly

Figure 9.4 Electronic or E-waste Is the Fastest-growing Source of Solid Waste. *Source*: Curtis Palmer. Wikimedia Commons. https://creativecommons.org/licenses/by/2.0/deed.en

disposed of, the toxic material inside the electronics is often released into soil, water, and air, potentially harming both people and the environment.

The release of toxic chemicals like polybrominated diphenyl ethers (PBDEs), and heavy metals like lead, chromium, and manganese from e-waste leads to many environmental and health issues. A 2021 study found that people living in e-waste-exposed regions had significantly elevated levels of heavy metals and persistent organic pollutants.[5] Children and pregnant women were especially vulnerable. Elevated toxic chemical levels had a negative impact on neonatal growth. Hormone level alterations were demonstrated in e-waste-exposed populations.

Solutions to the problems of e-wastes have been difficult to implement, although a variety of national and international efforts are in place.[6] There are several potential solutions:

- **Behavioral changes:** If individual users postpone upgrading electronics for as long as possible, it both reduces the need for new electronics and decreases the flow of older electronics into the e-waste stream.
- **Reuse:** If electronics are still in working order or require only minor repairs, they can often be reused by other users. Used electronics are returned to the manufacturer or donated to a charity. Refurbishing and upgrading older electronics provide inexpensive electronics for users who might otherwise be unable to afford them. Reuse also reduces the introduction of e-waste to the waste stream.

- **Recycling:** Used electronics that cannot be repaired or reused can be recycled at a dedicated e-waste recycling facility. The metals and chemicals in electronic waste are a resource for the manufacture of new electronic devices. Not only does recycling provide a use for used materials, but it also reduces the necessity to mine valuable resources from the earth.

While recycling is a promising solution to our e-waste problem, it has a long way to reach its potential. As noted above, only 17.4 percent of e-waste in 2019 was formally collected and recycled.[7] Much of the remainder becomes part of the "informal" recycling industry, which is particularly prevalent in developing countries. The *International Labour Organization* (ILO) and the WHO estimate that millions of women and child laborers working in the informal recycling sector around the world are at risk of e-waste exposure.[8]

9.2 SOLID WASTE MANAGEMENT

One of humanity's greatest modern challenges is determining how we are to deal with the billions of tons of solid waste that we generate every year. Earlier in this chapter we explored landfills for municipal solid waste and recycling for e-waste, but there are many other ways to deal with solid waste.

The concept of *solid waste management* generally refers to the collecting, treating, and disposing of solid waste. An entire multibillion-dollar industry arose in the United States and worldwide that provides waste management services for communities of all sizes. However, recent evaluations of our waste management strategies have compelled us to rethink how we deal with solid waste. One useful formulation arranges waste management strategies this way:

- **Waste Management**: Reduces the harm of solid waste by moving them away from populated areas but does not decrease the amounts.
- **Waste Reduction**: Emphasizes using less wasteful materials and focuses on reuse, recycling, and composting to decrease waste volume.
- **Integrated Waste Management (IWM):** Uses a variety of strategies. IWM is a relatively new approach to the sustainable management of solid wastes covering all stages of waste management, including generation, segregation, transfer, sorting, treatment, recovery, and disposal in an integrated manner, with an emphasis on maximizing resource use efficiency.

The IWM strategy is shown graphically in figure 9.5. IWM is a holistic waste prevention, recycling, composting, and disposal program. These systems consider methods to prevent, recycle, and manage solid waste in ways that

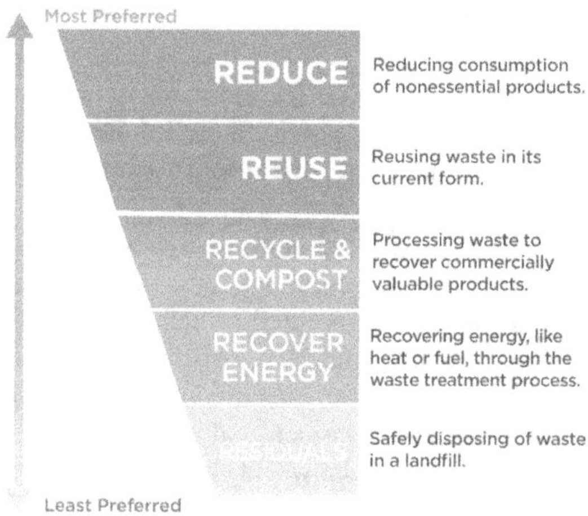

Figure 9.5 The Integrated Waste Management Hierarchy. City of Lincoln, NE. Reduce and Reuse. *Source*: https://www.lincoln.ne.gov/City/Departments/LTU/Utilities/Solid-Waste-Management/Reduce-and-Reuse (originally: http://www.epa.gov/garbage/faq.htm)

most effectively protect human health and the environment. IWM involves evaluating local needs and conditions, and then selecting and combining the most appropriate waste management activities for those conditions. The major ISWM activities are waste prevention, recycling and composting, and combustion and disposal in properly designed, constructed, and managed landfills. Each of these activities requires careful planning, financing, collection, and transport.

We can describe the components of IWM using the hierarchy from figure 9.5.

- **Reduce:** Source reduction, or "waste prevention," attempts to reduce the use of materials in the production of products in order to reduce waste at the consumer end. Source reduction strategies include using less packaging, designing products to last longer, and improving the efficiency of manufacturing. They help reduce handling, treatment, and disposal costs.
- **Reuse:** Reuse is the practice of using a material over and over in its current form. It preserves some or all of the energy and materials that went into producing an item in the first place. Societies around the world have long embraced the practice of reuse by finding alternate uses for an item rather than disposing or recycling it and reusing products and materials. Reuse of products can lead to substantial cost savings and are a staple in many developing countries.
- **Recycle:** Recycling and Composting. Recycling is a process that involves collecting, reprocessing, and/or recovering certain waste materials (e.g.,

glass, metal, plastics, paper) to make new materials or products. *Primary (closed-loop) recycling* is recycling into materials of the same type (e.g., aluminum from cans recycled into new aluminum cans). *Secondary recycling* is recycling materials into different products (e.g., plastic from bottles recycled into polyester for fabrics). *Tertiary recycling* involves the chemical alteration of a material in order to make it reusable for a different product (e.g., melting metals, chemically treating old paper, and breaking down plastic bottles to make brand new products). Some recycled organic materials are rich in nutrients and can be used to improve soils. The conversion of waste materials into soil additives is called *composting*. Recycling and composting generate many environmental and economic benefits. For example, they create jobs and income, supply valuable raw materials to industries, produce soil-enhancing compost, and reduce greenhouse gas emissions and the number of landfills and combustion facilities. Home composting has become very popular in many communities (figure 9.6).
- **Recover Energy:** Energy recovery, or "waste-to-energy" systems, are various technologies that convert solid waste into usable forms of energy including heat, fuels, and electricity. WTE can occur through a number of processes such as incineration, gasification, pyrolysis, anaerobic digestion, and landfill gas recovery. An example of a solid waste-to-energy facility is in figure 9.7.
- **Residuals:** This includes solid waste disposal, such as landfilling and combustion. These methods are used to manage waste that cannot be prevented or recycled. A common way to dispose of solid waste is to place it in properly designed, constructed, and managed landfills, which we described in section 9.1 (figure 9.3). An alternative to landfills is solid waste combustion, which is the controlled burning of waste, which helps reduce its volume. Properly designed and constructed, combustion facilities can be used as waste-to-energy facilities.

The IWM approach has gained considerable popularity in recent years, in no small part because of its emphasis on waste reduction strategies. The USEPA encourages federal agencies, states, and municipalities to adopt Integrated Waste Management Plans (IWMPs) as an efficient and cost-effective way to reduce open dumping, effectively manage solid waste, and protect human health and the environment. Several government agencies have adopted IWMPs, including the United States Army Environmental Command.[9] At the international level, the *United Nations Environment Programme* recommends IWMPs as a way to deal with solid waste in increasingly crowded conditions.[10]

Figure 9.6 A Backyard Composter. *Source*: Doc James. Own work. Wikimedia Commons. https://creativecommons.org/licenses/by-sa/4.0/deed.en

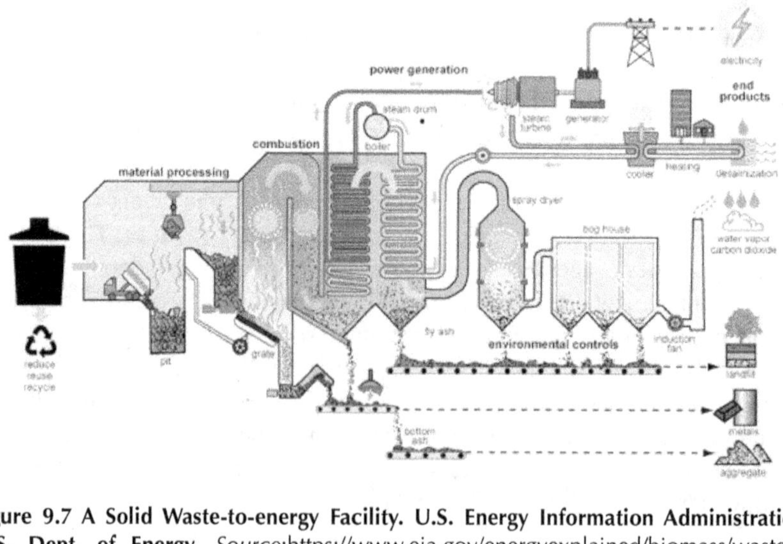

Figure 9.7 A Solid Waste-to-energy Facility. U.S. Energy Information Administration, U.S. Dept. of Energy. *Source*:https://www.eia.gov/energyexplained/biomass/waste-to-energy-in-depth.phption

9.3 THE REGULATION OF SOLID WASTE

Prior to the 1960s, solid wastes were not recognized as a significant health threat, and their disposal was largely unregulated. Solid wastes were simply buried or burned (with little regulation) or dumped into water bodies. There were few legal challenges, and most of those were based on common-law actions such as common law nuisance.

After World War II ended in the 1940s, Americans began to address pollution problems at home, including failures by other laws to regulate an increasing problem with solid waste disposal. In 1965, Congress passed the Solid Waste Disposal Act (SWDA), which was primarily a grant and research program that encouraged states to develop area-wide solid waste disposal plans.[11] The SWDA was amended in 1970 by the Resource Recovery Act, which authorized grants to state and local governments for "recovering materials and energy from solid wastes." However, these early laws were generally viewed as weak and ineffective, doing little to solve the health and environmental problems of accumulating solid and hazardous wastes.

By the 1970s, Americans were growing weary of regulatory attempts that did little to address a growing public awareness of environmental disasters. The most familiar of these was the environmental tragedy at Love Canal, just outside Niagara, New York. In the early 1900s, William T. Love designed a model community that included an electric power station that would generate electricity using water from a short canal, Love Canal, that connected to the Niagara River. However, Love's model community plan failed, and the canal was turned into a municipal and industrial chemical dumpsite with little or no oversight. In 1953, the Hooker Chemical Company, then the owners and operators of the property, covered the canal with earth and sold it to the City of Niagara for one dollar. Almost 100 homes and a school were built on the abandoned dump site (figure 9.8). But in the 1970s, a series of rainy years caused the buried waste chemicals to rise to the surface, causing a health catastrophe that included cancer, premature deaths, and birth defects. The Love Canal site was abandoned, and the chemicals were eventually removed.

In 1976, Congress passed a series of amendments to the SWDA called the RCRA). RCRA is technically a series of amendments to the SWDA of 1965, although RCRA stands as one of our most significant environmental laws. Under RCRA's solid waste control provision, the EPA awards federal grants to states that develop programs consistent with federal guidelines. RCRA regulates non-hazardous as well as hazardous solid wastes, contains a prohibition on open dumping, and includes financial and technical assistance to states and regional authorities to develop EPA-approved waste management plans.[12]

Figure 9.8 Love Canal as It Appeared in the Late 1970s. *Source*: USEPA. https://cumulis.epa.gov/supercpad/SiteProfiles/index.cfm?fuseaction=second.photovideoaudio&id=0201290

RCRA contains a series of important definitions, including a definition of solid waste:

> The term "solid waste" means any garbage, refuse, sludge from a waste treatment plant, water supply treatment plant, or air pollution control facility and other discarded material, including solid, liquid, semisolid, or contained gaseous material resulting from industrial, commercial, mining, and agricultural operations, and from community activities.[13]

In other words, "solid wastes" do not have to be solids under the federal definition but can include liquid or even gaseous wastes. We will explore this definition more fully below, including an investigation into wastes that are excluded from the definition of solid wastes.

A particularly troubling form of solid waste is *hazardous waste*. Hazardous waste is a waste with properties that make it dangerous or capable of having a harmful effect on human health or the environment. Hazardous waste is generated from many sources, ranging from industrial manufacturing process wastes to batteries and may come in many forms, including liquids, solids, gases, and sludges.[14]

What are hazardous wastes? Under RCRA, hazardous waste is a type of solid waste. Hazardous wastes are any "discarded" materials, whether they are actually solid, liquid, or gaseous in nature.

What makes solid waste hazardous? SWDA Section 7003(5) defines hazardous waste broadly as solid waste that may "cause, or significantly contribute to an increase in mortality or an increase in serious irreversible, or incapacitating reversible, illness; or . . . pose a substantial present or potential hazard to human health or the environment when improperly treated, stored, transported, or disposed of, or otherwise managed."[15]

Under RCRA, there are two ways that a waste can be determined to be hazardous. First, it is a hazardous waste if the EPA has placed it on its list of hazardous materials. These are called *listed hazardous wastes*. Second, if public or private testing determines that a waste is hazardous, but it does not appear on an EPA list, then it is called a *characteristic or identified hazardous waste*. Either way, a listed or characteristic hazardous waste is one that is hazardous because it is toxic, reacts violently with other chemicals, or is corrosive.

Hazardous wastes are further subdivided by the USEPA into three classes:

- **F Wastes**: from non-specific sources (e.g., used solvents, landfill leachate)
- **K Wastes**: from specific sources (e.g., chemical, pesticide, or pharmaceutical manufacturing); and
- **U and P Wastes**: toxic hazardous wastes (U Wastes) and acutely hazardous wastes (P Wastes).

The USEPA has provided a useful flow chart to help users identify hazardous wastes, as shown in figure 9.9.[16]

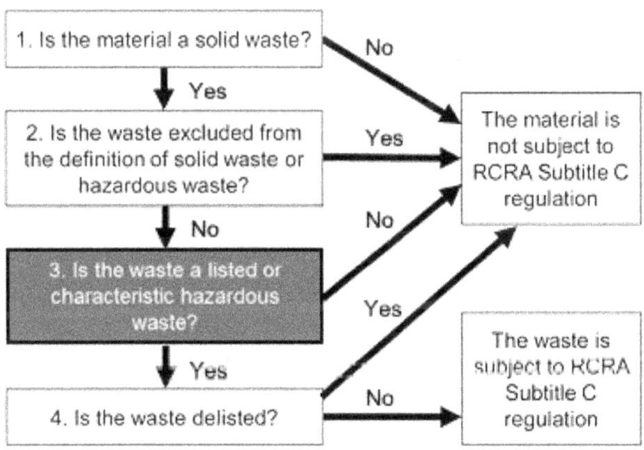

Figure 9.9 Steps in the Hazardous Waste Identification Process from the USEPA. *Source*: USEPA. Criteria for the Definition of Solid Waste and Solid and Hazardous Waste Exclusions. https://www.epa.gov/hw/criteria-definition-solid-waste-and-solid-and-hazardous-waste-exclusions.

The RCRA Cradle-to-Grave Regulatory System

The RCRA permit system is designed to provide a "cradle to grave" system of documentation identifying hazardous waste, and controlling its treatment, storage, and disposal.

Every facility that handles hazardous waste is classified and regulated separately. A *facility* is "any building, structure, installation, equipment . . . well, pit, pond, lagoon, impoundment, ditch, landfill . . . [or] any site or area where a hazardous substance has been deposited, stored, disposed of, or placed, or otherwise come to be located."[17]

Facilities are placed into any of three categories that are regulated differently:

1. **Generators**: Regulated so that no hazardous wastes are undetected;
2. **Transporters**: Regulated so that hazardous wastes are transported safely and identified as such; or
3. **Treatment, Storage, or Disposal (TSD) facilities**: Regulated to ensure that no hazardous wastes are released into the environment.

Let's investigate each of these categories.

Generators of Hazardous Waste

A *generator* of hazardous waste is any person who produces hazardous waste that is either listed or characteristic (identified). This very broad definition can include anything from a large industrial facility to a small chemistry classroom. Recognizing that generators produce waste in different quantities, the EPA established three categories of generators in the regulations: very small quantity generators, small quantity generators, and large quantity generators, each with increasingly complex regulations.

Each generator must have a plan for minimizing hazardous waste production and must keep extensive records of notices, manifests, and reports. Figure 9.10 is an example of the kind of documentation that must be maintained by academic institutions for science laboratories. When the wastes are transferred to a TSD facility, records must be maintained by the generator that track the path of the wastes. If a generator accumulates wastes that exceed specified limits, then it becomes a "storage facility" that is subject to additional regulatory requirements.

HAZARDOUS WASTE
Eastern Washington University
Environmental Health and Safety (EH&S) 509-359-2788

Chemical Composition	ml	%
isopropyl alcohol		70%
glycerol		15%
Water		10%
Trace Fat / Phenol		5%
Total		100

CORROSIVE REACTIVE NON-HAZARDOUS
✓ TOXIC ✓ IGNITABLE OXIDIZER
_ OTHER (EXPLAIN)

Date Filled 8/11/2020	Labeled By Awesome
Department BIOLOGY	Phone 1111
Building SCIENCE	Room 411
REPLACE CAP AFTER FILLING	CALL EH&S FOR PICKUP

Figure 9.10 An Example of the Kind of Documentation that Must Be Maintained by Small Generators of Hazardous Waste, in This Case a Laboratory at Eastern Washington University. *Source*: https://inside.ewu.edu/ehs/environmental-programs/hazardous-waste/information-for-waste-generators/

Transporters of Hazardous Waste

Hazardous waste *transporters* are individuals or entities that move hazardous waste from one site to another by highway, rail, water, or air. Transporters can be anything from tanker trucks to train cars to boats or even airplanes. They are a familiar sight to anyone who travels the interstate highway system in the United States

Because hazardous waste transporters move regulated wastes on public roads, highways, rails, and waterways, the USEPA and the U.S. Department of Transportation (USDOT) jointly developed the hazardous waste transporter regulations that apply to all hazardous waste transporters. Transporters are given identification numbers that allow for easier tracking of hazardous wastes by the USEPA, USDOT, and state and local agencies.

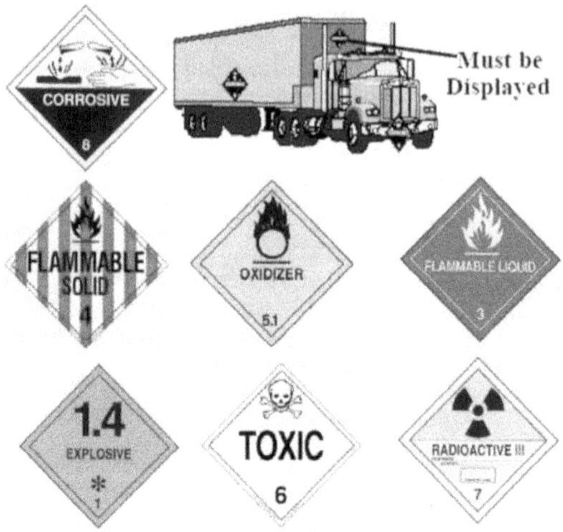

Figure 9.11 Examples of Placards Required to Be Displayed by Hazardous Waste Transporters. *Source*: Wikimedia Commons, Public domain work. https://creativecommons.org/publicdomain/zero/1.0/deed.en

Hazardous waste transporters must carry with them the manifest from the generator during transport, identify the waste, and deliver the manifest along with the waste to the ultimate destination. Transporters must also display placards that identify the nature of the wastes they transport, making it easier for first responders and local officials to identify the materials in case of an emergency and to respond appropriately. Figure 9.11 illustrates some of the placards that are displayed by transporters,

Treatment, Storage, and Disposal (TSD) Facilities

The final link in RCRA's cradle-to-grave system is the *TSD facility* that follows the generator and transporter in the chain of waste management activities. The regulations pertaining to TSD facilities are more stringent than those that apply to generators or transporters. They include general facility standards as well as unit-specific design and operating criteria. The general facility standards consist of good housekeeping provisions for any facility that handles hazardous waste. The unit-specific technical requirements are designed to prevent the release of hazardous waste into the environment. The regulations also address the different types of hazardous waste units available to treatment, storage, and disposal facilities.

Treatment, storage, and disposal activities differ and can be described this way:[18]

- **Treatment**–Uses processes such as incineration or oxidation to alter the character or composition of hazardous wastes. Some treatment processes enable waste to be recovered and reused, while other treatment processes dramatically reduce the amount of hazardous waste.
- **Storage**–Involves temporarily holding hazardous wastes until they can be treated or disposed of. Hazardous waste is commonly stored prior to treatment or disposal and must be stored in containers, tanks, containment buildings, drip pads, waste piles, or surface impoundments that comply with the RCRA) regulations.[19]
- **Disposal**–Involves permanently containing hazardous wastes. The most common type of disposal facility is a landfill, where hazardous wastes are disposed of in carefully constructed units designed to protect groundwater and surface water resources.

Owners and operators of TSD facilities are responsible for obtaining permits, and there are criminal sanctions for failing to do so. Operators are required to conduct *compliance monitoring* during the life of the facility, and for a period (often indefinite) until there are three consecutive years during which no groundwater protection standard has been exceeded. Compliance monitoring involves the collection of samples by a hydrogeologist for laboratory analysis, and inspections by state, local, and federal officials. The purpose of compliance monitoring is to assure that all TSD facilities meet federal guidelines for the protection of groundwater.[20]

Hazardous waste landfills are subject to special performance standards to ensure that they are designed to prevent leakage. Technological requirements must reflect all "improvements in the technology of control and measurement."[21]

RCRA Enforcement

The effectiveness of any law can be measured by its provisions for enforcement. Section 7003 of RCRA states that any time the past or present handling, storage, treatment, or disposal of solid or hazardous wastes "may present an imminent and substantial endangerment to health or the environment," the EPA is authorized to sue any past or present owner or operator of a generator, transporter, or TSD facility in federal District Court.[22] Persons acting in a corporate capacity may be individually liable if they are personally involved in the conduct that violates §7003.

RCRA Section 3008(d) authorizes the EPA or state agencies to use criminal judicial actions to enforce against people who knowingly violate RCRA, including RCRA's corrective action requirements.[23] Information on criminal enforcement is available on the EPA's criminal enforcement program website.[24]

RCRA Section 7002 contains the citizen suit provision. It authorizes citizens to bring enforcement actions against potential or actual violators of RCRA, including those who violate RCRA's corrective action requirements, and against the EPA in federal District Court.[25] Citizens must provide notice to the EPA, to the state in which the violation occurred, and to the alleged violator before any action can be taken.

Leaking Underground Storage Tanks (LUSTs)

Underground storage tanks have been used for over a century to store fuel, chemicals, pesticides, and many other liquid chemicals in what appeared to be very safe conditions. However, it was recognized in the 1960s that leakage from underground tanks that stored materials such as gasoline, kerosene, and fuel oil had begun to leak from the tanks. Abandoned gasoline service stations were a major culprit, but so were abandoned and/or razed houses and factories with underground fuel storage tanks. The greatest potential hazard from a LUST is that the petroleum or other hazardous substances can seep into the soil and contaminate groundwater, the source of drinking water for nearly half of all Americans. A LUST can present other health and environmental risks, including the potential for fire and explosion. Most of these LUSTs must be removed using dangerous extraction procedures (figure 9.12).

Figure 9.12 Underground Storage Tank Removal in Washington, D.C. *Source*: Wikimedia Commons, Public domain work. https:// creativecommons.org/publicdomain/zero/1.0/deed.en

RCRA Sections 9001–9008 contain the LUST program. The LUST program authorizes the USEPA to set national performance standards for buried tanks containing "regulated substances" (including all hazardous substances under SWDA and CERCLA).[26] Owners of tanks abandoned after 1973, or brought into operation since 1985, must send a notice, including the size, age, type, location, and use of the tank, to the appropriate state agency.

The definition of underground tanks excludes things such as residential septic tanks and any farm or residential tank less than 1,100 gallons used for fuel.

9.4 CERCLA AND HAZARDOUS WASTE REMEDIATION

Since the earliest days of the industrial revolution (and possibly as early as the Roman Empire), humans have been disposing of hazardous and toxic wastes by simply burying them, burning them, or disposing of them in water bodies. This past contamination has left a legacy of damage to human health and the environment.

The effects of exposure to hazardous substances on human health have been documented for decades. These exposures may irritate the skin or eyes, make it difficult to breathe, cause headaches and nausea, or result in other types of illness. Some hazardous substances can cause far more severe health effects, including cancer, genetic mutations, physiological malfunctions (e.g., reproductive impairment, kidney failure, etc.), and birth defects.[27] Modern medical research continues to explore these effects.[28]

Impacts on the environment can be equally devastating, although environmental damage does not receive as much attention as injury to human health. Scientific evidence suggests that exposure to hazardous waste can result in injury and death to organisms in a lake or river, destroy animals and plants in a contaminated area, cause major reproductive complications in animals, or otherwise limiting the ability of an ecosystem to survive. Certain hazardous substances also have the potential to explode or cause a fire, threatening both animal and human populations.

An excellent example of research on the effects of hazardous waste on wildlife is the research by Dr. Mike Hooper (Texas Tech University) and his colleagues, who used wildlife biomonitoring to determine whether contaminants at hazardous waste sites find their way into the food chain at biologically significant levels. His team studied prothonotary warblers at a hardwood bottomland floodplain (a seasonal swamp) that is on the National Priority List of contaminated sites. The team placed over 200 nest boxes throughout the study site (figure 9.13), and collected diet, egg, and nestling tissue samples. They demonstrated that warblers accumulated renal mercury from their diet

Figure 9.13 A Nest Box Used in Dr. Hooper's Study of Hazardous Substances in Food Chains. *Source*: National Institute of Environmental Health Science. Wildlife Biomonitoring at Hazardous Waste Sites. Superfund Research Program. https://www.niehs.nih.gov/research/supported/centers/srp/phi/archives/ecology/wildlife

at levels that correlated with contaminated soil and sediment levels, establishing that environmental contaminants find their way into the food chain and into the bodies of organisms in an ecosystem.

However, recognizing the problem is only the first step. Is it possible to clean up, or *remediate*, the past contamination? The estimated costs of a massive cleanup of all potentially dangerous chemical dumps in the United States are in the hundreds of billions of dollars.

Congress began to examine possible solutions to the remediation problem in the early 1970s. In response to environmental catastrophes like Love Canal (and the knowledge that existing environmental laws were inadequate to the task), considerable pressure was placed on Congress. Consequently, the CERCLA,[29] also known as the "Superfund" law, was passed in the waning days of the ninety-sixth Congress in 1980.

The passage of CERCLA was a milestone in U.S. environmental law. RCRA, the Toxic Substances Control Act (TSCA), the Federal Insecticide, Fungicide, and Rodenticide Act (FIFRA), the Clean Air Act, and the Clean Water Act all have provisions designed to regulate the production, distribution, and disposal of hazardous and chemical wastes, but none deal comprehensively with remediating past disposal problems.

CERCLA is somewhat unusual as an environmental law. It has no clear statement of goals and purposes, it lacks the usual complex system of regulation found in other environmental laws, and it relies instead on a few simple rules of liability. Reminiscent of the approach of NEPA, CERCLA relies on a form of "common law" to fill in crucial details. CERCLA is designed to pick up where other laws, primarily RCRA, leave off. In fact, CERCLA cannot be read separately from RCRA, with which it overlaps considerably.

CERCLA was extensively amended and reauthorized by the Superfund Amendments and Reauthorization Act of 1986 (SARA).[30] The SARA amendments created the "Superfund," which is a national trust fund, funded by a tax on the chemical industry, to pay for the cleanup of hazardous waste sites. SARA was passed over the objections of then-president Reagan and many members of Congress who felt that the cost was too large and that the charges to private RPs amounted to a value-added tax. SARA was designed to overcome the shortcomings of the original CERCLA. SARA is one of the longest and most intimidating of all environmental laws. We will discuss CERCLA and the SARA amendments together. In addition, the SARA amendments created the Emergency Planning and Community Right-to-Know Act (EPCRA), which we will discuss below.

Under CERCLA (and the SARA amendments), Congress gave the federal government broad authority to regulate hazardous substances, to respond to releases of hazardous substances, and to develop long-term solutions for the nation's most serious hazardous waste problems. It does this by incorporating several unusual and controversial approaches:

- CERCLA is *retroactive*. Under retroactive liability, a responsible party (RP) can be held liable for hazardous waste release occurring prior to 1980 (before CERCLA was enacted). In essence, this means that a responsible party can be held liable for an action that was not illegal at the time that it originally occurred.[31]
- CERCLA imposes *strict, joint, and several liability* on responsible parties. These are common law principles. Under *strict liability,* a person is legally responsible for the consequences of an action even in the absence of fault or criminal intent on the part of the defendant. *Joint and several liability* means that all of the defendants are responsible for the act and for the damages. The plaintiff, however, may seek to collect the awarded damages from all of the defendants (*joint liability*), or from some or only one (*several liability*) if they choose. If only one defendant ends up paying, they may seek reimbursement from the other defendants.
- CERCLA *prevents pre-enforcement reviews* of agency decisions. In other environmental laws, pre-enforcement reviews allow potentially responsible parties (PRPs) to challenge the action in court before it is

enforced against them. But under CERCLA, a responsible party cannot take legal action against an agency until after legal action is brought against them.[32]

The CERCLA Remediation System

In order to accomplish its ambitious goals, CERCLA uses four strategies:

1. An information-gathering and reporting system;
2. Authorization of federal responses to hazardous waste emergencies and the cleanup of dump sites;
3. The creation of a fund to pay for the cleanup; and
4. The imposition of strict, joint, and several liability on persons contributing to hazardous substances at inactive sites.

The system is complex, but surprisingly logical.

Information Gathering and Reporting

CERCLA Section 103(c) requires owners of inactive sites to notify the EPA of the existence of the site, the amount and types of any hazardous substances, and any known or likely releases of hazardous substances. States must then develop "inactive site inventories and priority lists" from this notification data.[33] The data submitted to the EPA from the states are then used to produce the Comprehensive Environmental Response, Compensation, and Liability Information System (CERCLIS), which is a database that lists each site for which the EPA received notice. Each site on the CERCLIS list is reviewed to determine whether there is "imminent and substantial danger" allowing the EPA to conduct a cleanup of the site.[34] EPA may place a lien on the property while investigating the contamination to prevent potentially responsible parties from escaping liability by selling or altering their property.

The information gathering process continues with the National Oil and Hazardous Substances Contingency Plan (NCP), which was developed by the EPA using notification information and other data. The NCP lists priorities among the known or threatened hazardous chemical release sites[35]

The next step in the process was to create the National Priorities List (NPL), which is a list of the sites on the CERCLIS list that require remedial action. The NPL is intended primarily to guide the EPA in determining which sites warrant further investigation, since the EPA only conducts remedial actions at sites on the NPL. Sites on the NPL are listed in order of priority based on "relative risk."[36] Some NPL rankings have been successfully challenged in court.[37] As of 2023, there are 1,340 NPL sites in the United States, with another thirty-nine sites proposed.[38]

Each site on the NPL list is assigned a number, or rank. The EPA developed the Hazard Ranking System (HRS) as part of its NCP, which is a tool to determine rank. The HRS assigns a numerical score for contaminated sites based on potential pathways for the migration of the contamination, toxicity of the contaminants, waste quantities, and nearby populations of people.[39]

After a site is listed on the National Priorities List (NPL), a remedial investigation/feasibility study (RI/FS) is performed at the site. The remedial investigation (RI) involves the collection of data to characterize site conditions, determine the nature of the waste, assess risk to human health and the environment, and conduct testing to evaluate the performance and cost of available treatment technologies. The feasibility study (FS) examines alternative remedial actions.

The next step in the process is the actual remediation, or removal, of hazardous substances. The EPA considers several factors in determining when and how the removal of hazardous materials will take place. These factors include actual or potential exposure to hazardous substances by nearby populations, animals, or food chains; contamination of drinking water supplies or ecosystems; materials near the surface that may migrate; weather conditions; and the threat of fire or explosions.[40] EPA then designates the procedures to be used in either a removal or a remedial action. The EPA categorizes response actions as either *removal actions* (emergency, time-critical, or non-time-critical) or *remedial actions* (generally longer-term response actions).

During its preliminary investigations, EPA uses the concept of Natural Resources Damage (NRD) to determine the appropriate type and amount of restoration needed to offset impacts to fisheries, wildlife, habitats, and human uses impacted by oil spills and hazardous waste sites.[41] During this process, Natural Resource Trustees use Natural Resource Damage Assessments (NRDAs) to assess injury to natural resources. NRDAs are used to calculate the monetary cost of restoring injuries to natural resources that result from releases of hazardous substances or discharges of oil.

CERCLA Responses

Once the EPA determines that remediation is necessary, a remedial plan is then prepared. Individuals (or entities) found to be responsible parties (RPs) may be permitted (or required) to assist in carrying out the remedial plan. However, RPs are prohibited from taking any action without the EPA's approval and approval of other RPs.

To a degree, CERCLA allows RPs to participate in the planning and remedial action for a site, known as *remedy selection*. The remedy that is

Figure 9.14 EPO Excavation during Cleanup of Agriculture Street Landfill Superfund Site in New Orleans, Louisiana. *Source*: USEPA. http://www.epa.gov/region6/6sf/louisian/ag_street/pictorials

eventually selected must protect human health and the environment, comply with applicable or relevant and appropriate requirements (ARARs), be cost-effective, and utilize permanent solutions available. Once EPA has selected the remedial process, a Record of Decision (ROD) is produced that explains which cleanup alternatives will be used. It contains information on site history, site description, enforcement activities, and the remedy selected for cleanup. The development of the ROD also includes consideration of future uses of the site. These documents give both guidance and security to RPs. Construction and landscape professionals (figure 9.14) conduct the actual cleanup.

CERCLA Cost Recovery

The costs of CERCLA cleanups is often measured in millions of dollars. One of the goals of CERCLA is to provide a fund (the "Superfund") that would finance the massive cleanup costs for the thousands of hazardous waste sites across the nation.[42] To replenish the fund (and prevent bankruptcy of the nation), CERCLA contains provisions for recovering cleanup costs from RPs.

CERCLA Section 107 allows the government or private citizens to recover costs incurred in a cleanup from RPs. Since liability is strict, joint, and several, it is one of the most feared (and most litigated) aspects of environmental law. Section 107 establishes the system of strict, joint, and several liability for all Potentially Responsible Parties (PRPs) for a release of hazardous substances.

The idea is to make it possible to sue everybody who might be liable.[43]

The CERCLA compensation scheme seems to make sense. The parties responsible for the pollution ("RPs"), who presumably profited from it, are held financially responsible for the cleanup.

However, liability may be placed disproportionately and unfairly on a later RP. This has led to a great deal of CERCLA litigation.

Any person who is, or has in the past, operated a "facility" is an RP. A problem is how to treat corporate officers and shareholders of corporations that operate facilities (or have done so in the past). In most areas of business, corporate officers are not personally liable even if they were personally negligent. Under CERCLA, courts have held liable both corporate officers and majority shareholders if they personally participated or had actual control. Minority shareholders and officers that do not participate are generally not liable.

What if the responsible corporation has dissolved or was replaced by a successor corporation? Courts have generally held that dissolved corporations are liable if they participated in the contamination. Where the issue is the liability of parent corporations over their subsidiaries, courts usually look at the amount of control the parent exercised. Corporations that purchase assets of a corporation requiring cleanup may be liable for costs under the *continuity of enterprise* principle.

A common situation involves several RPs, each of whom has made a substantial contribution to the contamination. How do you decide who pays what? Most courts hold that all RPs are subject to joint and several liability for cleanup costs without respect to degree of fault (CERCLA §§107 and 113(f)(1)), unless defendants can prove a reasonable basis for apportioning the costs. CERCLA §113 authorizes courts to consider equitable factors such as a polluter's relative culpability in deciding how much each party will have to contribute.[44]

CERCLA Defenses

Defenses by RPs against paying cleanup costs are difficult given CERCLA's broad scope. Under CERCLA, cost recovery does not take place for naturally occurring substances, materials within residential or commercial buildings, or materials released into drinking water systems through ordinary use.[45] These materials are either not regulated by federal law or are regulated by laws other than CERCLA.

Because CERCLA demands strict liability, it doesn't matter if a defendant did everything possible to prevent contamination. One of the few defenses

Figure 9.15 Under CERCLA §107(b), Property Owners Are Usually Not Liable for this Kind of Contamination. *Source*: USEPA. Public domain image. https://en.wikipedia.org/wiki/United_States_Environmental_Protection_Agency

listed by CERCLA is that the contamination resulted from an act of God.[46] However, some acts of God, such as floods and lightning, may not be defenses in areas where they occur frequently.

Another defense is when a release is the result of an "act or omission of a third party."[47] This defense has worked when a complete outsider (who delivers the contamination by "midnight dumping" or the like) was responsible for the contamination (figure 9.15).

Still another defense is the "innocent landowner" defense. The innocent landowner defense is usually an effective defense if a landowner is completely innocent and used due care to discover the contamination. To qualify, a landowner must have undertaken at the time of acquisition all appropriate inquiry into the previous ownership consistent with customary practices. The contiguous property owner's defense protects innocent adjacent property owners.

The Emergency Planning and Community Right-to-Know Act (EPCRA)

EPCRA is the Emergency Planning & Community Right-to-Know Act. Enacted in 1986, EPCRA is also known as Title III of the Superfund Amendments and Reauthorization Act of 1986 (SARA) (figure 9.16). EPCRA is designed to provide assistance to local communities in protecting public health, safety, and the environment from chemical hazards.

Figure 9.16 EPCRA Is Title III of the SARA Amendments to CERCLA. *Source*: Public domain document. USEPA. Minnesota Department of Public Safety, https://dps.mn.gov/divisions/hsem/epcra/Pages/default.aspx.

To implement EPCRA, Congress requires each state to appoint a State Emergency Response Commission (SERC). The SERCs are required to divide their states into Emergency Planning Districts and to name a Local Emergency Planning Committee (LEPC) for each district. EPCRA Sections 311 and 312 require the annual submission of chemical inventory data (known as Tier II reports) by facilities to state and local planning officials for incorporation into ongoing emergency planning.[48] Among other things, Section 311 requires that the owner or operator of a facility must submit a Material Safety Data Sheet (MSDS) for each hazardous chemical which meets or exceeds a specified threshold quantity at the facility.[49]

EPCRA Section 313 requires facilities to quantify and submit annual releases of certain chemicals for incorporation into a national database known as the Toxics Release Inventory (TRI). The TRI is a resource for learning about toxic chemical releases and pollution prevention activities reported by industrial and federal facilities. TRI data support informed decision-making by communities, government agencies, companies, and others.

The Special Problems of Brownfields

Part of our legacy of terrestrial pollution is the problem of what have come to be called *brownfields*. Brownfields are areas where past contamination of land means that new establishments will not relocate to a contaminated site because of potential liability under CERCLA. Rather than building on contaminated brownfields (and incurring CERCLA liability), many industrial and commercial establishments choose to locate to uncontaminated "greenfields." These brownfields often leave a legacy of decreased property values, blighted areas that invite crime, and contamination that poses a health risk to local residents and ecosystems.

A familiar type of brownfield is an abandoned industrial site where hazardous chemicals were stored under conditions where they have leaked chemicals into surrounding soils and groundwater. Abandoned gasoline stations are frequent brownfield locations (figure 9.17). Urban properties often contain brownfields because past uses of the property left a residue of contamination.

There are several possible solutions to the brownfields problem. One is to allow responses that remediate the contaminated property to less restrictive uses, like "industrial" or "commercial" instead of "residential" so that the level of cleanup is not as rigorous. For example, figure 9.17 shows an abandoned service station "brownfield" in New Jersey that was replaced with a fire station, which required less cleanup than a school or hospital would.

The EPA introduced a series of Brownfields Initiatives in 1995. One major initiative included delisting over 25,000 of the least contaminated sites from the NPL, thus releasing owners from potential CERCLA liability. Another initiative was the preparation of a new Lender Liability Guidance Document,

Figure 9.17 Abandoned Service Station Built in Trenton, NJ was Replaced by a Fire Station. *Source*: New Jersey Depertement of Environmental Protection. Trenton, NJ: The Site of a National Brownfield Success!. https://www.nj.gov/dep/srp/brownfields/news/trenon0306.htm

which was not a regulation but to give potential brownfield purchasers some comfort.

Still another initiative was a series of municipal exemptions from CERCLA liability. These were designed to encourage municipalities to acquire urban brownfields either through purchase or through involuntary acquisition such as foreclosure and eminent domain. Several major cities have taken advantage of these programs. Municipal exemptions have allowed the City of Chicago to acquire, remediate, and then develop contaminated sites on the south side of Chicago, where former industries had seriously contaminated the area.

* * *

Review Questions

1. List and describe the six types of solid waste.
2. Describe the disposal of household solid waste.
3. Why is electronic waste, or "e-waste," especially dangerous?
4. What are the steps in modern solid waste management?
5. Describe RCRA's cradle-to-grave system for hazardous waste management.
6. Describe the CERCLA information gathering system.
7. CERCLA imposes strict, joint and several liability on responsible parties. What does this mean?
8. Explain the process of cost recovery under CERCLA and its relationship to the Superfund.
9. What defenses are available to PRPs to avoid CERCLA liability?
10. What are brownfields, and why are they a potential problem for communities?

Case Analysis

Since the early 1900s, the Anaconda Copper Smelter in Butte, Montana, contaminated an area of over 300 square miles with arsenic and lead. Beginning in the 1980s, Atlantic Richfield, the owners of the smelter, then an NPL site, worked with the EPA to implement a cleanup plan under CERCLA for a remediation expected to continue through 2025. During that time, a group of ninety-eight landowners sued Atlantic Richfield in Montana state court for common law nuisance, trespass, and strict liability, seeking restoration damages. Under Montana law, all money recovered has to be spent on property rehabilitation, even though the amount they received far exceeds the recovery costs as estimated by the EPA. However, EPA determined that the landowners were also potentially responsible parties (PRPs), since some contamination may have come from them. This raised an interesting legal issue. Under CERCLA §122(e)(6), PRPs are prohibited from taking remedial action without EPA approval, which had not been granted. The Montana Supreme Court held that the common-law case could proceed because the landowners were not PRPs, and that decision was appealed to the U.S. Supreme Court.

The following is an excerpt from the U.S. Supreme Court's decision, written by Chief Justice Roberts.[50] The discussions of jurisdiction and references to the dissenting opinions have been removed.

> Although the Montana Supreme Court answered the jurisdictional question correctly, the Court erred by holding that the landowners were not potentially responsible parties under the Act and therefore did not need EPA approval to take remedial action. Section 122(e)(6), titled "Inconsistent response action," provides that "[w]hen either the President, or a potentially responsible party . . . has initiated a remedial investigation and feasibility study for a particular facility under this chapter, no potentially responsible party may undertake any remedial action at the facility unless such remedial action has been authorized by the President." 42 U.S.C. §9622(e)(6). Both parties agree that this provision would require the landowners to obtain EPA approval for their restoration plan if the landowners qualify as potentially responsible parties.
>
> To determine who is a potentially responsible party, we look to the list of "covered persons" in [CERCLA] §107, the liability section of the Act, §9607(a). "Section 107(a) lists four classes of potentially responsible persons (PRPs) and provides that they 'shall be liable' for, among other things, 'all costs of removal or remedial action incurred by the United States Government'" (§9607(a)(4)(A)). The first category under §107(a) includes any "owner" of "a facility." §9607(a)(1). "Facility" is defined to include "any site or area where a hazardous substance has been deposited, stored, disposed of, or placed, or otherwise come to be located." §9601(9)(B). Arsenic and lead are hazardous substances. 40 CFR §302.4, Table 302.4. Because those pollutants have "come to be located" on the landowners' properties, the landowners are potentially responsible parties.
>
> The landowners . . . argue that even if the landowners were once potentially responsible parties, they are no longer because the Act's six-year limitations period for recovery of remedial costs has run, and thus they could not be held liable in a hypothetical lawsuit. 42 U.S.C. §9613(g)(2)(B).
>
> This argument collapses status as a potentially responsible party with liability for the payment of response costs. A property owner can be a potentially responsible party even if he is no longer subject to suit in court. As we have said, "[E]ven parties not responsible for contamination may fall within the broad definitions of PRPs in §§107(a)(1)–(4)." That includes "'innocent' . . . landowner[s] whose land has been contaminated by another," who would be shielded from liability by the Act's so-called "innocent landowner" or "third party" defense

in §107(b)(3). Ibid. See also 42 U. S. C. §9607(b)(3). The same principle holds true for parties that face no liability because of the Act's limitations period.

Interpreting "potentially responsible parties" to include owners of polluted property reflects the Act's objective to develop, as its name suggests, a "Comprehensive Environmental Response" to hazardous waste pollution. Section 122(e)(6) is one of several tools in the Act that ensure the careful development of a single EPA-led cleanup effort rather than tens of thousands of competing individual ones.

Yet under the landowners' interpretation, property owners would be free to dig up arsenic-infected soil and build trenches to redirect lead-contaminated groundwater without even notifying EPA, so long as they have not been used within six years of commencement of the cleanup. We doubt Congress provided such a fragile remedy for such a serious problem. And we suspect most other landowners would not be too pleased if Congress required EPA to sue each and every one of them just to ensure an orderly cleanup of toxic waste in their neighborhood. A straight-forward reading of the text avoids such anomalies.

. . .

The judgment of the Montana Supreme Court is . . . vacated in part. The case is remanded for further proceedings not inconsistent with this opinion.

It is so ordered.

Questions

1. The landowners were never notified by the EPA that they were responsible for any cleanup costs, and were no longer PRPs. Were they right?
2. The landowners apparently felt that after the six-year limitations period for recovery of remedial costs had run, they were no longer PRPs and should be allowed to begin their cleanup. On what basis did the Supreme Court disagree?
3. Is the property owned by the landowners a "facility" under CERCLA §9601(9)(B)? What was the Court's reasoning for saying that they are?
4. Does this decision make sense? Should a private property owner be denied the ability to clean up their property just because they happen to fall into a category where they might someday be held responsible for a tiny part of the contamination?

NOTES

1. Solid Waste. In: *Compendium of WHO and Other UN Guidance on Health and Environment.* World Health Organization, 2021. WHO/HEP/ECH/EHD/21.02.

2. The USEPS website is an excellent source of information on wastes of all kinds, with an introduction to the regulatory framework. https://www.epa.gov/report-environment/wastes.

3. 29 CFR §1910.1030.

4. Forti, V., Baldé, C.P., Kuehr, R., & Bel, G. 2020. *The Global E-Waste Monitor 2020*. United Nations University and United Nations Institute for Training and Research.

5. Parvez, S.M., Jahan, F., Brune, M.N., Gorman, J.F., Rahman, M.J., Carpenter, D., Islam, Z., Rahman, M., Aich, N., Knibbs, L.D., & Sly, P.D. Health Consequences of Exposure to e-Waste: An Updated Systematic Review. *The Lancet Planet Health, 5*(12), e905–e920. https://doi.org/10.1016/S2542-5196(21)00263-1.

6. The USEPA website discusses issues of e-waste pollution and regulatory solutions: https://www.epa.gov/international-cooperation/cleaning-electronic-waste-e-waste.

7. Forti, V., Baldé, C.P., Kuehr, R., & Bel, G. 2020. *The Global E-Waste Monitor 2020*. United Nations University and United Nations Institute for Training and Research.

8. ILO/WHO. 2021. *Child Labour: Global Estimates 2020, Trends and the Road Forward.* International Labour Organization. https://www.ilo.org/ipec/Childlabour statisticsSIMPOC/lang--en/index.htm.

9. U.S. Army Environment Command. *Integrated Solid Waste Management (ISWM).* https://aec.army.mil/protect/ISWM.

10. UNEP. *Guidance on Policy and Legislation for Integrated Waste Management during a Pandemic.* https://www.unep.org/resources/publication/guidance-policy-and-legislation-integrated-waste-management-during-pandemic.

11. Solid Waste Disposal Act of 1965, 42 U.S.C. § 6901-6992.

12. 42 U.S.C. § 6901 et seq.

13. *42 U.S.C. §2011 et seq.*

14. USEPA. Learn the Basics of Hazardous Waste. https://www.epa.gov/hw/learn-basics-hazardous-waste#hwid.

15. 42 U.S. Code § 6973.

16. The USEPA website has a useful set of links to help users find their way through the hazardous waste regulatory system. This includes a discussion of the "cradle to grave" system discussed in Section 9.6. https://www.epa.gov/hw.

17. RCRA §101(9), 42 U.S.C. § 9601(9).

18. USEPA. *Hazardous Waste Management Facilities and Units.* https://www.epa.gov/hwpermitting/hazardous-waste-management-facilities-and-units.

19. 40 CFR parts 264-265.

20. See USEPA. *Hazardous Waste Compliance Monitoring.* https://www.epa.gov/compliance/hazardous-waste-compliance-monitoring.

21. 42 U.S.C. §3004(o).

22. 42 U.S.C. §6973.

23. 42 U.S.C. §6928.

24. USEPA. *Criminal Enforcement.* https://www.epa.gov/enforcement/criminal-enforcement.

25. 42 U.S.C. §6972.

26. 42 U.S.C. §6991. The USEPA LUST program is described at https://www.epa.gov/ust.

27. See USEPA. *Health and Ecological Hazards Caused by Hazardous Substances*. https://www.epa.gov/emergency-response/health-and-ecological-hazards-caused-hazardous-substances.

28. Public health studies have been particularly valuable tools. For example, see: Fazzo, L., Manno, V., Iavarone, I., Minelli, G., De Santis, M., Beccaloni, E., Scaini, F., Miotto, E., Airoma, D., & Comba, P. (2023). The Health Impact of Hazardous Waste Landfills and Illegal Dumps Contaminated Sites: An Epidemiological Study at Ecological Level in Italian Region. *Frontiers in Public Health, 11* 996960. https://doi.org/ 10.3389/fpubh.2023.996960.

29. 42 U.S.C. §9601 et seq.

30. 42 U.S.C. §11001 et seq.

31. In 1996, a federal district court in Alabama broke with precedent and held that CERCLA's liability provisions cannot be applied retroactively (*United States v. Olin Corp.*, 927 F. Supp. 1502 (S.D. Ala. 1996)). Subsequent courts have not followed this decision, but we have probably not heard the last of the issue.

32. The unavailability of pre-enforcement reviews under CERCLA has been controversial. See the thoughtful discussion by Mark W. Schneider in 2016 at https://acoel.org/revisiting-cercla-pre-enforcement-review/.

33. 42 U.S.C. §9603.

34. 42 U.S.C. §9606.

35. 42 U.S.C. §9605(a)(b)(B).

36. (42 U.S.C. §9605(a)(8)(A)).

37. See, for example, *Atlantic Richfield Co. v. Christian*, 140 S. Ct. 1335 (2020).

38. This and other useful information is at USEPA's website Superfund: National Priorities List (NPL) at https://www.epa.gov/superfund/superfund-national-priorities-list-npl.

39. The factors used by EPA in the HRS system are discussed on its web site, USEPA. *Hazard Ranking System (HRS).* https://www.epa.gov/superfund/hazard-ranking-system-hrs.

40. 40 CFR §300.65(b)(2).

41. The National Oceanic and Atmospheric Administration (NOAA) describes the process at: https://oceanservice.noaa.gov/facts/nrda.html.

42. CERCLA §111 and §112 deal with the uses of the fund and procedures for making claims against it. 42 U.S.C. §§9601-9602.

43. 42 U.S. Code §9607.

44. 42 U.S. Code §9613.

45. 42 U.S. Code §9604(a)(3).

46. 42 U.S. Code §9607(b).

47. 42 U.S. Code §9607(b).

48. 42 U.S.C. §§11021-11022.

49. USEPA, General reporting requirements and thresholds for EPCRA 311. https://www.epa.gov/epcra/general-reporting-requirements-and-thresholds-epcra-311.

50. Atlantic Richfield Co. v. Christian, 590 U.S.___,140 S.Ct. 1335 (2020).

Chapter 10

Energy, Environment, and Law

Chapter Learning Outcomes

At the end of this chapter, the reader will be able to:

- *Contrast renewable and non-renewable sources of energy.*
- *Compare the costs and benefits of coal, petroleum, and natural gas as energy sources.*
- *List the costs and benefits of the use of nuclear power.*
- *Describe the advantages and disadvantages of renewable energy sources.*
- *Identify the sources of legal regulation of energy production.*

One of the key drivers of the success of the United States as a global superpower is its effective national energy system. Its "power grid" is made up of thousands of both public and private energy producers and transmitters and is regulated by hundreds of federal, state, and local agencies. Unfortunately, the success of our energy system brings with it a series of health and environmental challenges. In this chapter, we will examine our energy system and its impact on health and the environment and examine the energy regulatory framework.

10.1 ENERGY USE WORLDWIDE

Energy systems have been central to the development of human societies around the world, although the consumption of energy such as electricity is not uniform across the world. As shown in figure 10.1, developing countries consume far less energy per person than developed countries. Energy systems are probably the most important single determinant of growth of industrial

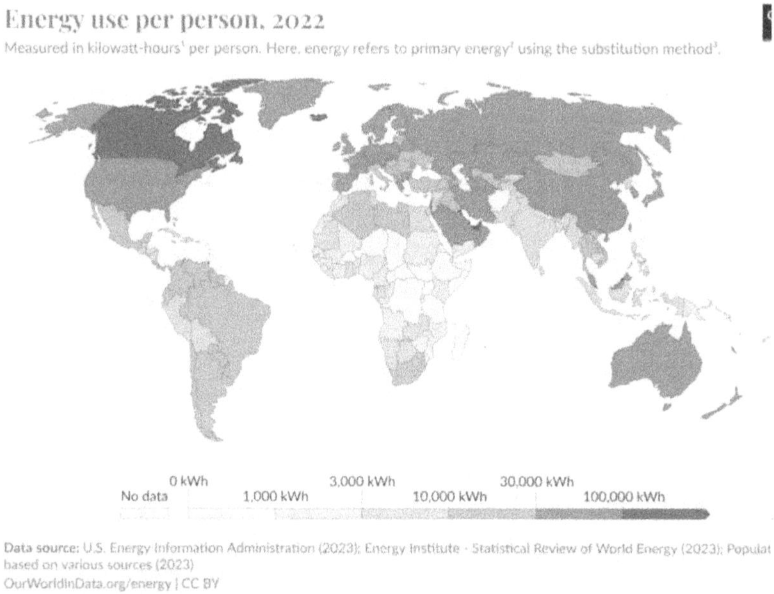

Figure 10.1 Consumption of Energy such as Electricity Is much Lower in Developing Countries across the World. *Source*: Wikimedia Commons. Our World in Data. https://creativecommons.org/licenses/by/3.0/deed.en

societies and modern economies. Unfortunately, they have also resulted in many negative environmental developments. Current energy systems are the primary source of carbon dioxide (CO_2) emissions, accounting for 84 percent of total global CO_2 emissions and 64 percent of global greenhouse gas (GHG) emissions related to human activities.[1] Past trends suggest that this percentage will likely increase in the future, especially if our energy needs continue to be met by fossil fuels.

Let's begin by examining the sources of our energy. In discussing the sources of energy for consumption, a good place to start is by looking at renewable versus non-renewable energy sources. In general, *non-renewable energy* comes from finite sources that cannot be regenerated within a human lifetime and will eventually run out, like oil, coal, and natural gas. By contrast, *renewable energy* comes from unlimited, naturally replenished resources like the sun, wind, and water. Both energy sources can be used for electricity generation, heating and cooling, and transportation.

We will start by examining non-renewable and renewable energy sources in detail, including an investigation of their effects on health and the environment. We will then examine the regulatory framework and look to the future.

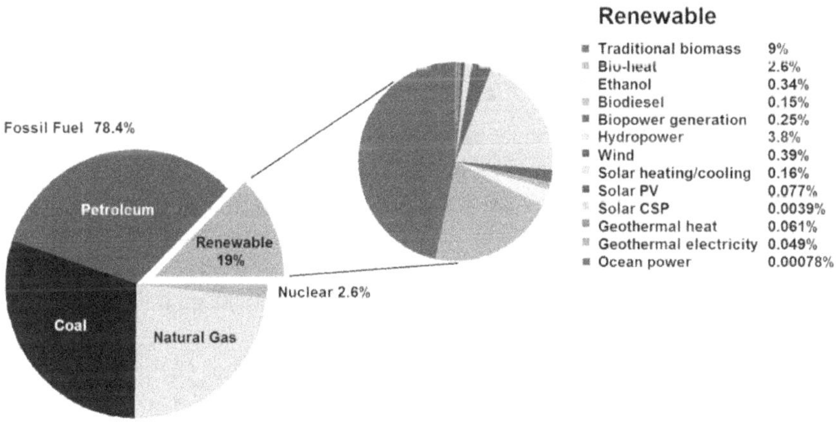

Figure 10.2 Total World Energy Consumption by Source (2013). *Source*: Wikimedia Commons. Delphi234. Own work. https://creativecommons.org/licenses/by/3.0/deed.en

Figure 10.2 illustrates the relative consumption of non-renewable and renewable fuels worldwide. It is clear that the vast majority of the world's energy consumption is from non-renewable fossil fuels. We will further examine both renewable and non-renewable energy sources.

10.2 NON-RENEWABLE ENERGY SOURCES

Non-renewable energy comes from sources that will run out and are not replenished in the foreseeable future. Most non-renewable energy sources are *fossil fuels*, meaning that they are formed by natural processes over millions of years in which organic material from dead plants and animals was gradually converted by high temperature and pressure into the material that we find today. The major fossil fuels are coal, petroleum (oil), and natural gas.

Nuclear fuel (uranium) is usually included as a non-renewable fuel. Uranium forms over billions of years during violent galactic processes. Uranium is consumed when it is used, so it is not "renewed" in the usual sense.

Carbon is the main element in most fossil fuels. The time period during which fossil fuels formed (about 360–300 million years ago) is called the Carboniferous Period. Burning fossil fuels releases the carbon stored in the fuels as carbon dioxide (CO_2) and carbon monoxide (CO).

There are both advantages and disadvantages to the different sources of non-renewable energy. Let's explore these energy sources in more detail.

Coal

There are several forms of coal: *anthracite, bituminous, subbituminous, and lignite*. The ranking depends on the types and amounts of carbon the coal contains and the amount of heat energy the coal can produce.

Coal is the most abundant and cheapest energy resource in the world. Coal is burned in furnaces of various types. It can be burned as a form of heat in buildings, but it is used mostly to produce electricity and steel. Coal is very energy-dense compared to other energy sources: one gallon of American gasoline corresponds approximately to 40 acres of plant matter in terms of energy.

Worldwide coal reserves are mainly in the United States (25 percent), with Russia in second place (16 percent), and China third (12 percent). The United States has the largest coal resources on earth. Figure 10.3 illustrates by country where coal reserves are found.[2] The United States has coal reserves that could last up to 250 years, but an increase in consumption of just 4 percent per year could reduce the reserves to just sixty-four years.

Of course, there are major downsides to the use of coal as an energy source. First, mining and transportation are expensive and dangerous, as illustrated in figure 10.4. Second, burning coal produces CO_2, SO_2, and contributes to air pollution, acid rain, and climate change. As we discussed in chapter 6, there are technologies to reduce these emissions, but they are expensive, and their availability is limited in some countries.

Figure 10.3 Total World Reserves of Coal by Country (2020). *Source*: Wikimedia Commons. Our World in Data. https://creativecommons.org/licenses/by/3.0/deed.en

Figure 10.4 Coal Mining Is Expensive and Dangerous. Left is surface mining in Jharia, India. Right is underground mining in Westland, New Zealand. *Source*: Wikimedia Commons. https://creativecommons.org/licenses/by/2.0/deed.en

Petroleum

Petroleum is a liquid or a semi-solid and differs in appearance depending on the geology of its formation. The retrieval and refining processes are expensive and hazardous, and shipping can be expensive and dangerous. Its advantage over coal is that is can be transported by pipeline under some circumstances. There are many problems associated with petroleum use, including oil spills and air pollution, especially CO_2 and other greenhouse gases.

Petroleum is a thick, flammable, yellow-to-black mixture of gaseous, liquid, and solid hydrocarbons that occurs naturally beneath the earth's surface. Petroleum can be separated into fractions including natural gas, gasoline, naphtha, kerosene, fuel and lubricating oils, paraffin wax, and asphalt. It is used as a raw material for a wide variety of derivative products.

As illustrated in figure 10.5, the world's greatest oil reserves are located in the Middle East, Venezuela, North America, and Russia.

Oil from petroleum is an extremely versatile fuel; it is transformed into many other products. Oil-derived fuels (such as gasoline) burn easily, burn hot, and the current transportation industry is designed to use them nearly exclusively.

Extraction of underground petroleum is an expensive and dangerous activity. Drilling "rigs" are constructed both on land and in the ocean to dig into oil reservoirs and search for new oil deposits (figure 10.6). Working these rigs can be very hazardous; safety is a major concern since any petroleum found is under pressure and highly flammable.

Land-based rigs are smaller and more numerous than offshore rigs. Smaller crews are needed to operate them (usually 5–8 workers) and they are fairly easily moved from place to place. Offshore rigs are much larger and more expensive to build. Larger crews are needed to operate offshore rigs (usually 25–40 workers) and they are extremely hard to move.

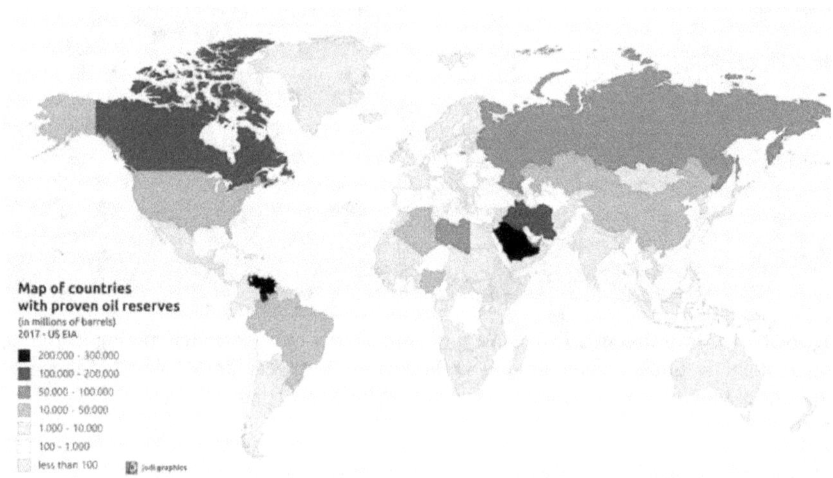

Figure 10.5 Map of Countries with Proven Oil Reserves According to the U.S. Energy Information Agency (2017). *Source*: Wikimedia Commons. Jo Di Graphics. Own work. Wikimedia Commons. https://creativecommons.org/licenses/by/2.0/deed.en

Figure 10.6 Oil Drilling Rigs Are Constructed on Land (left) and in the Ocean (right). *Source*: Wikimedia Commons. https://creativecommons.org/licenses/by/2.0/deed.en

In order to extract gasoline and other useful products from petroleum, complex refineries are necessary (figure 10.7). Refineries are expensive to build and operate, and require large, specialized workforces and equipment. They consume large tracts of land and produce water and air pollution.

Despite these expenses and dangers and its production of air pollution, petroleum remains a major energy source all over the world because it is one of the most available and least expensive energy sources.

However, our dependence on petroleum as an energy source comes with a large environmental price. Burning petroleum produces air pollution (especially SO_2, CO_2, and other greenhouse gases), and the extraction and refining of petroleum damage ecosystems. Unintentional releases of petroleum, often

Figure 10.7 A Pertroleum Refinery. *Source*: Wikimedia Commons. https://creativecommons.org/licenses/by/2.0/deed.en

called "oil spills," damage ecosystems, wildlife, and human health. The Deepwater Horizon spill in 2010 resulted in the release of over four million barrels of oil into the Gulf of Mexico, causing eleven deaths and an environmental catastrophe that continues today.[3]

Thousands of oil spills happen every year worldwide, both on land and in water. Environmental damage is most severe in the most sensitive ecosystems, so wetlands, coral reefs, and beach ecosystems are at the greatest risk. On land, an oil spill can damage surface structures and can pollute groundwater. Fortunately, the technology for remediating oil spills has improved markedly in the past thirty years.[4]

Natural Gas

There is evidence that natural gas produces less air pollution than coal or petroleum, but it also suffers from high production and transportation costs. Natural gas can be mined from deposits in the earth (including "fracking," discussed below), or it is a byproduct of petroleum refining. Some natural gas is produced by the breakdown of organic waste in landfills, as we discussed in chapter 9.

Natural gas is a fossil fuel, formed in a manner similar to petroleum. It is removed from the earth by drilling, and the pressure is usually sufficient to allow it to flow to the surface. The most common natural gas is methane, a colorless, odorless gas with a simple chemical structure (CH_4). Since it is colorless and odorless, a scent is added to it for safety purposes (a mercaptan), giving natural gas a sulfur-based, "rotten-egg" smell that cannot be removed. Approximately 24 percent of the world's energy is derived by burning natural gas.

An increasingly common and controversial form of natural gas recovery is unconventional natural gas extraction, or *fracking*, which is a process to extract gas from rock, especially shale. With this method, water, sand, and a mix of chemicals are forced under pressure into horizontally drilled wells, causing the shale to crack and release natural gas or oil. The process creates many small cracks in the rock and keeps them open so that gas can flow from the well. The mechanism is illustrated in figure 10.8.

Fracking has been used successfully for decades, but in recent years it has become controversial. Water quality is a concern because the hydraulic fracturing fluids used to fracture rock formations have chemicals that could harm human health and the environment, especially if they enter drinking water supplies. Other concerns relate to chemicals that are recovered and disposed of as wastewater. Emissions of air pollutants are another potential hazard.[5]

Three decades since the rapid expansion of fracking, researchers are revealing the scope of associated health effects and related costs. In some areas, such as Colorado, Oklahoma, Pennsylvania, and Texas, oil and natural

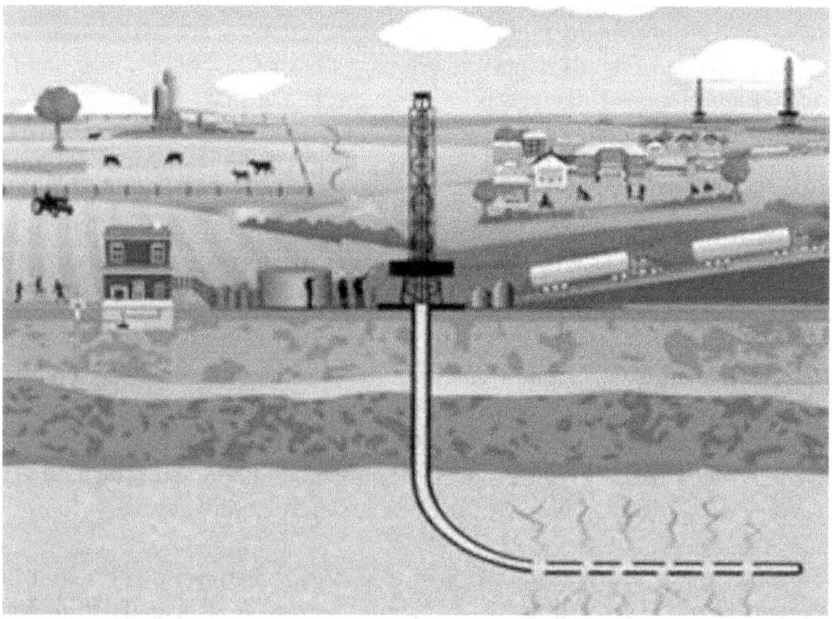

Figure 10.8 A Facility for Recovering Natural Gas by Hydraulic Fracturing, or Fracking. *Source*: National Institute of Environmental Health Sciences. Hydraulic Fracturing & Health. https://www.niehs.nih.gov/health/topics/agents/fracking

gas extraction and development may occur close to residential communities. Research has provided evidence demonstrating how fracking affects the health of people in nearby communities.[6]

Nuclear Energy (Uranium)

The radioactive element uranium can be used to generate electrical energy in power plants. The nuclei of uranium atoms are split apart into smaller nuclei in the process of nuclear fission. This process releases a tremendous amount of energy from just a small amount of uranium. However, the spent uranium fuel remains highly radioactive for thousands of years, making nuclear waste disposal a complex and controversial process.

The total supply of uranium in the world is very limited, making nuclear energy a non-renewable resource. Figure 10.9 illustrates worldwide uranium reserves in 2010.

Nuclear fuel has the distinct advantage of producing almost no air pollution. However, the mining and extraction processes are expensive and dangerous. When uranium ore is mined, extensive low-level tailings are left behind, which can damage ecosystems for many years. The next step in the process is milling and enrichment, which increase the proportion of U-235 by removing U-238. These steps leave still more radioactive tailings.

Nuclear power plants are expensive to build and maintain. Nuclear waste materials are extremely dangerous and difficult to store. Horrific accidents at several nuclear facilities have left nuclear energy with a huge public acceptance problem. These accidents at nuclear power plants have the potential to release large amounts of harmful radiation into the environment. The most

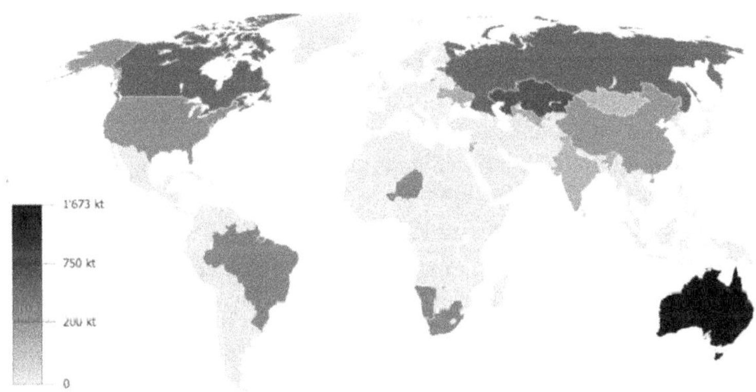

Figure 10.9 World Uranium Reserves in 2010. *Source*: Emilfaro. Own work. World Nuclear Association, Public Domain. https://commons.wikimedia.org/w/index.php?curid=11603907

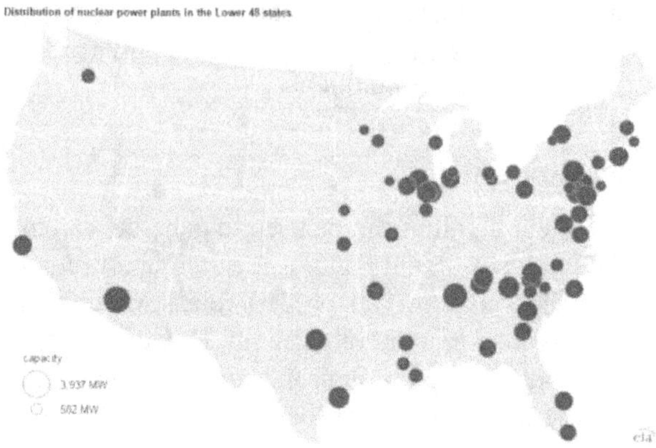

Figure 10.10 Map of the Locations of Nuclear Ppower Plants in the Lower Forty-eight States in 2017. The size of the dot indicates the capacity in megawatts. U.S. Energy Information Administration. Today in Energy. *Source*: https://www.eia.gov/todayinenergy/detail.php?id=30972

famous of these incidents were at the Chernobyl facility in Ukraine in 1986 and the Fukushima facility in Japan in 2010. Both accidents led to the loss of human lives, long-term health consequences for local populations, and environmental damage.

Despite these drawbacks, nuclear power plants continue to operate. Figure 10.10 illustrates the locations and capacities of nuclear power plants in the lower forty-eight states in 2017.

10.3 RENEWABLE ENERGY SOURCES

Renewable energy is energy that is collected from renewable sources, which are naturally replenished on a human timescale, such as sunlight, wind, rain, tides, waves, and geothermal heat. Renewable energy provides energy in important areas, including electricity generation, transportation, and energy services for rural (off-grid) areas.

Accessibility to renewable energy has increased in the past few decades. The technologies necessary to implement non-renewable energy sources have improved dramatically as scientific research has improved our understanding of renewable energy processes.[7]

As we will see, there are both advantages and disadvantages to renewable energy.[8] Some of the advantages include:[9]

- Renewable energy won't run out.
- Renewable energy has lower maintenance requirements.

- Renewable energy has numerous environmental benefits.
- Renewable energy decreases reliance on foreign energy sources.
- Renewable energy leads to cleaner water and air.

Some disadvantages of renewable energy include:

- Renewable energy has high upfront costs.
- Renewable energy is often not available at all times.
- Renewable energy sources have geographic limitations.
- Renewables aren't always 100 percent carbon-free.

We will explore solar energy, wind energy, water energy, biomass energy, and geothermal energy as renewable energy sources. Consider the value of each of these.

Solar Energy

Humans have known for millennia that we can warm ourselves or our homes by orienting ourselves to take advantage of the radiant energy of the sun. Simple technologies available today allow us to heat water and our homes by exposing liquid-filled panels to be warmed by the sun (figure 10.11). These are inexpensive and extremely efficient systems.

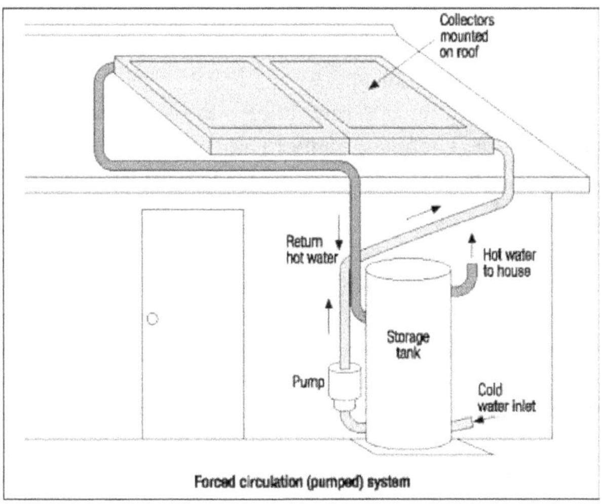

Figure 10.11 An Open Loop Solar Water Heating System. Water is pumped into the collectors on the roof during the day, where the water is heated by the sun and drained to the storage tank at night. These systems produce hot water and also heat the house. *Source*: Wikimedia Commons. Government of Western Australia. Public domain document.

Beginning in the twentieth century, we developed new technologies that store radiant energy from the sun in cells that convert the sun's energy into electricity that we can use as an energy source. This is called *photovoltaic (PV) solar energy*. Photovoltaic technologies, more commonly known as solar panels, generate power by using devices that absorb energy from sunlight and convert it into electrical energy through semiconducting materials. These devices, known as *solar cells*, are connected to form larger power-generating units known as modules or *solar panels*. These solar panels have no moving parts and can operate for an indefinite time without wearing out.

The process of converting solar energy from the sun into electrical energy depends on location and time of day. On a clear day, the amount of sunlight striking a square meter of earth is approximately equal to 1,000 watts. Photovoltaic solar systems offer many advantages, such as sustainable energy production, cost efficiency, flexibility, and independence from electricity suppliers. However, there are also disadvantages, such as weather dependence, initial cost, space requirements, and dependence on sunlight.

There are also concerns for the proper disposal of solar panels. Several hazardous chemicals are used in the manufacture of photovoltaic cells and panels, which must be carefully handled to avoid releasing them into the environment. Some types of PV cell technologies use heavy metals, and these types of cells and PV panels may require special handling when they are discarded. Some solar thermal systems use potentially hazardous fluids to transfer heat, and leaks of these materials could be harmful to the environment.[10] As we saw in chapter 9, several U.S. environmental laws regulate the use and disposal of hazardous materials.

Despite the drawbacks, solar energy is clearly seen as a path forward for renewable energy. Public and private solar energy "farms" are increasingly common (figure 10.12). There are many incentives, both nationally and internationally, to convert to solar energy. In the United States a purchaser of a solar energy system is eligible for the Solar Investment Tax Credit, or ITC, which provides a tax credit to the purchaser. Many states and municipalities offer separate tax credits or other incentives for conversion to solar energy.[11] But we have a long way to go. As of 2024, only about 5 percent of U.S. electricity is generated by solar energy systems.

Wind Energy

We learned in chapter 3 that wind is created by the uneven heating of the earth's surface, so it could be considered a solar energy source. Since wind is in plentiful supply, wind energy is a sustainable energy source that will be available as long as the sun's rays heat the planet.

Figure 10.12 A Public Solar Energy Farm Operated under the Bureau of Land Management's Solar Energy Project. U.S. Bureau of Land Management. Solar Energy. *Source*: https://www.blm.gov/programs/energy-and-minerals/renewable-energy/solar-energy

Figure 10.13 A Wind Turbine "Farm" in Southern California. *Source*: Wikimedia Commons. Erik Wilde, harvesting winds. https://creativecommons.org/licenses/by-sa/2.0/deed.en

Humans have harnessed wind as an energy source for millennia, most notably as a way to propel boats and ships over the water. Over the past few centuries, we have used *wind turbines*, or mechanisms operated by blades that move with the wind, in the form of *windmills* that draw water from underground sources.

More recently, we have adapted the concept of a wind turbine to create electricity. These wind turbines are often arranged into large "farms" on land or water where multiple turbines can provide electricity for entire communities (figure 10.13).

Predictably, there are both advantages and disadvantages to wind as a renewable energy source.[12] Advantages include the fact that wind is a renewable energy source. Wind turbines do not release emissions that can pollute the air or water (with rare exceptions), and they do not require water for cooling. Wind turbines may also reduce electricity generation from fossil fuels, which results in lower total air pollution and carbon dioxide emissions. Also, individual wind turbines have a relatively small physical footprint. Groups of wind turbines, sometimes called wind farms, are located on open land, on mountain ridges, or offshore in lakes or the ocean.

There are disadvantages to solar energy as well. Modern wind turbines can be very large machines that can be visually unappealing and noisy. Maintenance of wind turbines is difficult and expensive. There is evidence that birds and bats are injured or killed by turbine blades. These deaths may contribute to declines in the population of species also affected by other human-related impacts.

Of course, wind turbines are manufactured products, so they require the use of metals and other materials used to make wind turbine components. These have impacts on the environment as well, and fossil fuels may be used to produce the materials. Although most of the materials used to make wind turbines can be reused or recycled, turbine blades, as most are currently constructed, cannot be recycled.[13]

An innovative solution to some of the problems with traditional horizontal axis wind turbines (HAWT) like those in figure 10.13 is the use of vertical axis wind turbines (VAWT) like those in figure 10.14. With the VAWT configuration, there is no need to shift the direction of the unit to face the wind. There is also growing evidence that at least some VAWTs are more efficient, with lower maintenance costs and fewer environmental impacts than the HAWTs currently in use. VAWTs have been used successfully for decades in parts of Europe and are evidence that efficient, renewable energy sources are possible.

Water Energy (Hydropower)

Energy from water is among the most efficient and renewable sources available to us. Most hydropower involves flowing water that spins a turbine to produce energy in the form of electricity. For this reason, we often refer to it as *bioelectric* energy.

Hydropower depends on flowing water, which is a direct result of the water cycle that we discussed in chapter 3. Humans have used the force of flowing water in streams and rivers to produce mechanical energy for centuries. Mills that use hydropower are still used today throughout the world as an energy source to grind grain and operate machinery. Hydropower was one of the first

Figure 10.14 An Innovative Vertical Axis Wind Turbine. Energy.gov. How Do Wind Turbines Work? *Source*: https://www.energy.gov/eere/wind/how-do-wind-turbines-work

sources of energy used for electricity generation, and until 2019, hydropower was the leading source of total annual U.S. renewable electricity generation.

Hydroelectricity accounted for about 6.2 percent of total U.S. utility-scale electricity generation and 28.7 percent of total utility-scale renewable electricity generation. Hydroelectricity generation varies annually, and its share of total U.S. electricity generation generally decreased from the 1950s through 2020, mainly because of increases in electricity generation from other sources. Hydroelectricity's percentage share of total annual U.S. electricity generation from 2001 through 2022 averaged about 6.7 percent.[14]

The principle behind the operation of a *conventional hydroelectric facility* is simple. Water from a higher location (a reservoir) is pulled by gravity through a turbine that produces electrical current, to a river or stream at a lower elevation (figure 10.15).

In a *run-of-the-river system*, the force of the river's current applies pressure on a turbine without the need for a dam. The facilities may have a weir in the water course to divert water flow to hydroelectric turbines.

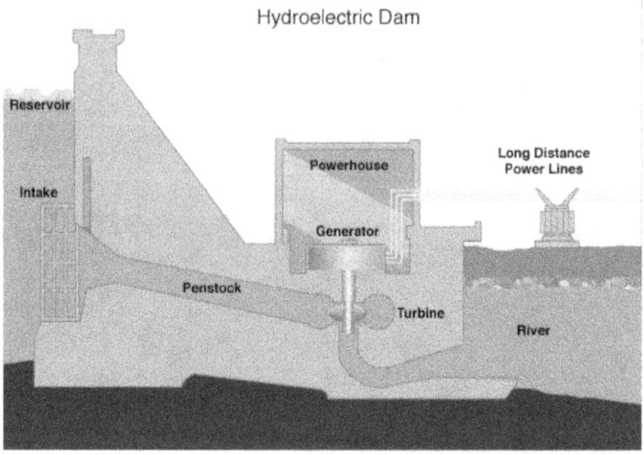

Figure 10.15 Diagram of a Dam and a Conventional Hydroelectric Power Plant. *Source*: Wikimedia Commons. Tennessee Valley Authority. https://creativecommons.org/licenses/by/3.0/deed.en

Another version of the hydroelectric facility is a *pumped-storage facility*. At times of low electrical demand, water is pumped into a higher reservoir. When the demand becomes greater, water is released back into the lower reservoir through a turbine. Pumped-storage hydroelectric systems generally use more electricity to pump water to the upper water storage reservoirs than they produce with the stored water, so they are not considered to be a renewable energy resource.

Hydroelectricity is an extremely popular energy source in locations with the proper topography and hydrology. There are about 1,450 conventional and 40 pumped-storage hydroelectric power plants operating in the United States (figure 10.16). They account for about 44 percent of U.S. electrical generating capacity.

Hydroelectric facilities are a common form of renewable energy throughout the world. It is estimated that there are at least 3,700 large hydroelectric power plants with a capacity of more than 1 MW worldwide, and an additional 82,800 smaller plants currently operating, under construction, or being planned.[15]

The largest hydroelectric dam in the world is the Three Gorges Dam on the Yangtze River in Hubei province, China. The Three Gorges Dam is the world's largest power station in terms of installed capacity (22,500 MW). In 2014, the dam generated 98.8 terawatt-hours (TWh) and held the world record but was surpassed by the Itaipú Dam, which set the new world record in 2016, producing 103.1 TWh.[16]

There are both advantages and disadvantages to hydroelectric energy.[17] Among the advantages are that hydroelectric energy is a renewable energy

Figure 10.16 Distribution of Conventional Hydroelectric Facilities in the Lower Forty-eight Sstates. *Source*: U.S. Energy Information Administration. 2017. Hydropower explained. https://www.eia.gov/todayinenergy/detail.php?id=30312oelectric generators are among the United States' oldest power plants.

source, it pairs well with other renewable energy sources, it is inexpensive in the long run, it is produced domestically, and the reservoirs can be used for irrigation.

There are also serious disadvantages to hydroelectric facilities. The major disadvantage is that damming rivers to create reservoirs often causes catastrophic damage to local ecosystems, in effect converting an existing ecosystem to a lake and interfering with fish migration.[18] In addition, they are expensive to build and maintain, reservoirs occupy large areas of land that cannot be used for anything else, and they can only be constructed where local topography and hydrology permit.

Biomass Energy (Bioenergy)

Biomass energy, or bioenergy, is energy from renewable organic material from plants and animals. Biomass contains stored chemical energy from the sun that is produced by plants through photosynthesis. Biomass can be burned directly for heat or converted to liquid and gaseous fuels through various processes. Biomass comes from biological materials such as wood, straw, sawdust, animal wastes, or recycled paper. It has been used as a source of energy since humans first learned to control fire.

Biomass is the term used for organic material that has stored sunlight as chemical energy. It is one of the largest renewable energy sources in use today. Biomass is usually classified as primary biomass, or biomass harvested directly for energy, or as secondary biomass, biomass from residues and waste.[19]

Biomass was the largest source of total annual U.S. energy consumption until the mid-1800s. In 2022, biomass accounted for about 5 percent of U.S.

total primary energy consumption. Today, biomass is used for heating and electricity generation and as a transportation fuel. In many countries, biomass is an important fuel, especially for cooking and heating in developing countries.[20]

Biomass sources for energy used today include:[21]

- *Wood and wood processing waste*—firewood, wood pellets, wood chips, lumber and furniture mill sawdust and waste, and waste material from pulp and paper mills.
- *Agricultural crops and waste materials*—corn, soybeans, sugar cane, switchgrass, woody plants, algae, and crop and food processing residues—are mostly to produce biofuels.
- *Biogenic materials in municipal solid waste*—paper products; cotton and wool products; and food, yard, and wood wastes.
- *Animal manure and human sewage* for producing *biogas* (renewable natural gas).

For most uses, biomass fuels must be converted into a usable form for that particular use. These fuels can be renewable replacements for fuels derived from non-renewable sources. These are the major biomass conversion techniques used today:[22]

- *Direct combustion* is the most common method for converting biomass to useful energy. Any form of biomass can be burned to heat buildings or water, and to generate electricity in steam turbines.
- *Thermochemical conversion* of biomass includes pyrolysis and gasification, which involve thermal decomposition processes in which biomass materials are heated in closed, pressurized gasifiers at high temperatures. The processes mainly differ in the temperatures and in the amount of oxygen present during the conversion.
- *Pyrolysis* entails heating organic materials to between 800°F and 900°F (400°C and 500°C) in the nearly complete absence of free oxygen. Biomass pyrolysis produces fuels such as charcoal, bio-oil, renewable diesel, methane, and hydrogen.
- *Gasification* entails heating organic materials to between 1,400°F and 1,700°F (800°C and 900°C) with injections of controlled amounts of free oxygen or steam into the vessel to produce a carbon monoxide- and hydrogen-rich gas called synthesis gas or *syngas*. Syngas can be used as a fuel for diesel engines, for heating, and for generating electricity in gas turbines. It can also be treated to separate the hydrogen from the gas, and the hydrogen can be burned or used in fuel cells. The syngas can be further processed to produce liquid fuels.

- *Hydrotreating* is used to process bio-oil (produced by fast pyrolysis) with hydrogen under elevated temperatures and pressures in the presence of a catalyst to produce renewable diesel, renewable gasoline, and renewable jet fuel.
- A chemical conversion process known as *transesterification* is used for converting vegetable oils, animal fats, and greases into fatty acid methyl esters (FAME) to produce biodiesel.

Biological conversion of biomass includes fermentation to make ethanol and anaerobic digestion to produce biogas. Ethanol is used as a vehicle fuel. *Biogas*, also known as biomethane or renewable natural gas, is produced in anaerobic digesters at sewage treatment plants and at dairy and livestock operations. It also forms in and may be captured from solid waste landfills, as we discussed in chapter 9. Properly treated renewable natural gas has the same uses as fossil fuel natural gas.

Biomass energy only accounts for about 5 percent of the energy budget of the United States,[23] but there are reasons to be optimistic. Figure 10.17 shows biofuel production by world region as of 2022. Worldwide biofuel production has increased 20-fold since 1990. The United States is clearly the world's leader in biofuel production, and there is every reason to believe that the trend will continue.

On the international stage, the United Nations Environment Programme has demonstrated strong support for the use of biofuels as a way to provide power, especially to developing countries, and to protect the environment.[24]

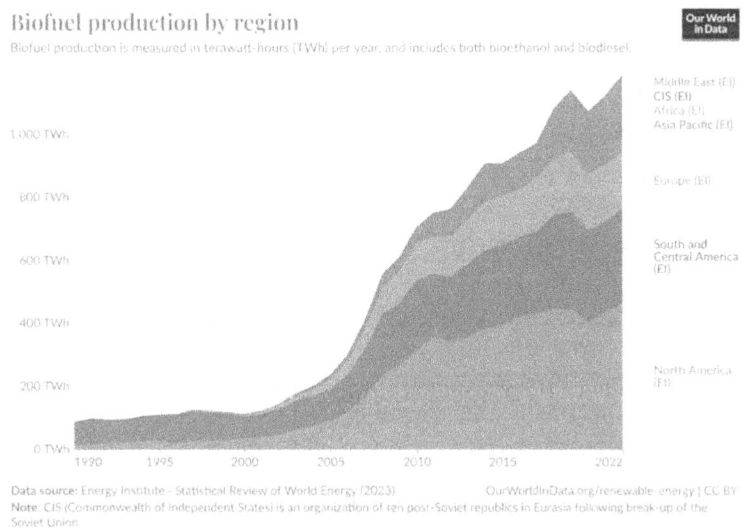

Figure 10.17 Biofuel Production by World Region as of 2022. *Source*: Wikimedia Commons. Our World in Data. https://creativecommons.org/licenses/by/3.0/deed.en

Geothermal Energy

Geothermal energy is a renewable energy source that comes from very hot materials deep within the earth. This natural heat arises from the ancient heat remaining in earth's core, from the decay of radioactive elements, and from friction where continental plates slide beneath each other. The amount of geothermal energy is massive. It is estimated that just 1 percent of the heat contained in the uppermost 10 kilometre of earth's crust is equivalent to 500 times the energy contained in all of the earth's oil and gas resources. However, this heat energy is not distributed evenly, seldom concentrated, and often at depths too great to be economically exploited.[25]

Geothermal energy is one of the oldest energy sources used by humans. Archaeological evidence suggests that the earliest direct use of geothermal power occurred at least 10,000 years ago in North America, where indigenous peoples used natural hot springs for both spiritual and practical purposes. In the first century CE, Romans conquered *Aquae Sulis*, now Bath, in England, and used the hot springs there to supply public baths and underfloor heating. The admission fees for these baths probably represent the first commercial use of geothermal energy. The world's oldest geothermal district heating system, in Chaudes-Aigues, France, has been operating since the fifteenth century. The earliest industrial exploitation began in 1827 with the use of geyser steam to extract boric acid from volcanic mud in Larderello, Italy.[26] Today, many hot springs are still used for bathing, and many people believe the hot, mineral-rich waters have health benefits.

Geothermal electricity generation requires water or steam at high temperatures (300°F to 700°F). Geothermal power plants are generally built where geothermal reservoirs are located, within a mile or two of the earth's surface. Geothermal electricity generation requires water or steam at high temperatures (300°F to 700°F). Geothermal power plants are generally built where geothermal reservoirs are located, within a mile or two of the earth's surface.

The modern use of geothermal energy is mainly either geothermal heat pumps or conversion to electricity. Direct geothermal heat pumps use the constant temperatures of rocks or water near the surface of the earth to heat and cool buildings. Geothermal heat pumps transfer heat from the ground (or water) into buildings during the winter and reverse the process in the summer.

Geothermal electricity generation requires water or steam at high temperatures (300°F to 700°F). The steam is used to drive turbines that generate electricity. Geothermal power plants are built where geothermal reservoirs are located within a mile or two of the earth's surface. The United States leads the world in geothermal electricity generation. In 2022, the United States had geothermal power plants in seven states, which produced about 0.4 percent (17 billion kilowatt-hours, kWh) of total U.S. utility-scale electricity generation. Utility-scale power plants have at least 1,000 kilowatts (or one megawatt) of electricity generation capacity. Figure 10.18 shows the countries with significant geothermal energy production.

In 2021, twenty-seven countries, including the United States, generated about 92 billion kilowatt-hours of electricity from geothermal energy. Indonesia was the second-highest geothermal electricity producer after the United States at about 16 billion kilowatt-hours—which was about 5 percent of Indonesia's total electricity generation. Kenya was the eighth-highest geothermal electricity producer, at about 5 billion kilowatt-hours, which was equal to about 43 percent of Kenya's annual electricity generation. Kenya had the largest percentage share of electricity generation from geothermal energy among all countries with geothermal power plants.[27]

There are several important advantages to the use of geothermal energy. It is a reliable source of energy and is not dependent on factors such as time and weather. It produces few, if any pollutants. Geothermal energy facilities have a relatively small land footprint, and the infrastructure tends to have greater longevity than other energy sources. It is adaptable for large and small-scale installations.

There are also disadvantages to geothermal energy. The most obvious is that it is very location-dependent. With today's technology, geothermal energy is only available in areas that are geologically active. In addition, there

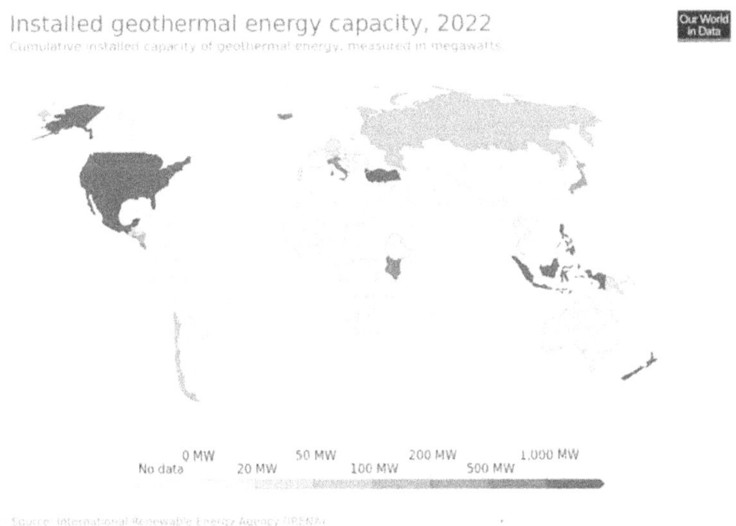

Figure 10.18 Installed Capacity of Geothermal Energy across the World. *Source*: Wikimedia Commons. Our World in Data. https://ourworldindata.org/grapher/installed-geothermal-capacity, CC BY 4.0, https://commons.wikimedia.org/w/index.php?curid=130881563

are high initial costs for geothermal energy facilities, and the extraction of subsurface hot water can lead to instability on the surface.[28]

Despite the drawbacks, geothermal energy has the potential to be an important source of renewable energy for the future.

In this section, we have seen that renewable energy sources are not only available, but they are also an important source of energy for the world. They can reduce the costs and availability of energy and dramatically decrease our dependence on fossil fuels that contribute to climate change.

10.4 THE REGULATION OF ENERGY

Prior to the 1920s, the role of the federal government in energy regulation was restricted to the mining and transport of oil, gas, and coal on federal lands. During the Great Depression and into the years of World War II, the federal government began its first attempts to establish a fragmented regulatory framework involving participation by many agencies.[29] As we have seen, federal agencies at that time were extremely mission-oriented, with an emphasis on supporting industry and little attention to health or the environment.

Among the earliest attempts at energy regulation was the Mineral Leasing Act of 1920,[30] which is the major federal law that regulates the leasing of public lands for developing deposits of hydrocarbons and other minerals. Previous to this act, these materials were subject to mining claims under the General Mining Act of 1872.

Regulation of the energy industry in the United States had its origin with the Federal Power Act (FPA)[31] of 1920, which created the Federal Power Commission. Major amendments to the FPA in 1935 and 1986 first allowed a national regulatory framework to develop. In 1946, the Atomic Energy Act[32] was passed in response to the Manhattan Project of the 1940s.

The Public Utility Regulatory Policies Act (PURPA)[33] was passed in 1978 in order to encourage, among other things, the conservation of electric energy, increased efficiency in the use of facilities and resources by electric utilities, the provision of equitable retail rates for electric consumers, the expeditious development of hydroelectric potential at existing small dams, and to promote the conservation of natural gas while ensuring that rates to natural gas consumers remain equitable.

One of the ways PURPA set out to accomplish its goals was through the establishment of a new class of generating facilities that would receive special rate and regulatory treatment. Generating facilities in this group are known as qualifying facilities (QFs) and fall into two categories: qualifying small power production facilities and qualifying cogeneration facilities.

PURPA has been modified several times during its history, but it remains an important part of U.S. energy regulation.

Regulatory development was slow in the middle years of the twentieth century. In 1977, the Federal Energy Regulatory Commission (FERC) was established within the newly created Department of Energy. The FERC assumed the regulatory functions of several agencies, including the Federal Power Commission. FERC was established as an independent regulatory agency that oversees the natural gas, oil, and electricity markets in the U.S. FERC regulates the transmission and sale of these energies (except the sale of oil), provides licenses for hydroelectric plants, and responds to certain environmental issues. The U.S. Nuclear Regulatory Commission (NRC) regulates the use of nuclear power and its use as a defense weaponry.

Energy regulation in the twenty-first century has been quite active, with attempts to address climate change and to increase market competition in order to serve the goal of cheap, reliable energy. The latter goal is often called "energy deregulation," but it has done little to reduce regulation of energy industries. The Energy Policy Act of 2005 changed the energy policy of the United States by providing tax incentives and loan guarantees for energy production.[34] The Act made several changes to U.S. energy policy. It changed daylight saving time by four to five weeks, depending on the year. It directs a study for the development of oil shale and tar sands resources on public lands, especially in Colorado, Utah, and Wyoming. It sets federal reliability standards regulating the electrical grid (in response to the Blackout of 2003). It created a number of tax credits favored by energy industries, such as the Nonbusiness Energy Property Tax Credit, but it did not include provisions for petroleum drilling in the Arctic National Wildlife Refuge (ANWR).

In 2007, the Energy Independence and Security Act (EISA)[35] was passed to encourage an increase in fuel economy standards for passenger vehicles, and the establishment of the first efficiency standard for medium-duty and heavy-duty commercial vehicles. In 2009, the American Recovery and Reinvestment Act (ARRA)[36] was a stimulus package that supported loans and investments into green energy technology, provided funding for an electric smart grid, created and modified renewable energy tax cuts, and provided funds to weatherize modest-income homes.

The 2021 Infrastructure Investment and Jobs Act[37] expanded funding for DOE research and development into carbon capture and storage, hydrogen economy tech, and grid battery storage. It promoted smart grid deployment and resiliency, and extended tax breaks to the nuclear power industry.

In 2022, the CHIPS and Science Act/Inflation Reduction Act[38] expanded funding for DOE research, development, and national security, extended and modified renewable energy tax cuts, weatherized low-income homes, and provided tax incentives for electric vehicles and new appliances.

These and other laws and regulations have created our modern energy regulatory system. Under the modern system, energy regulators promote investment in grid development and operation by providing clear and stable regulatory frameworks, setting appropriate incentives and tariffs, encouraging innovation, and facilitating interconnection and access. As we will see in chapter 11, energy regulation is an important tool in our programs for tackling climate change.

* * *

Review Questions

1. Describe the differences between renewable and non-renewable sources of energy.
2. Describe the types of non-renewable energy, and how they are acquired.
3. Explain the advantages and disadvantages of fossil fuel use.
4. Discuss fracking and its advantages and disadvantages.
5. Describe the types of renewable energy, and how they are acquired.
6. Explain the advantages and disadvantages of renewable energy sources.
7. Describe the history of the U.S. energy regulatory framework.
8. Identify the primary federal laws that apply to energy sources and their use in the United States.

Case Analysis

The regulatory environment for energy regulation is extremely complex. It is often difficult for individual parties to navigate the regulatory morass, so plaintiffs often resort to common-law actions against power companies or the government to challenge decisions related to energy. An interesting case of this kind occurred in Oklahoma in 2016 when a group of property owners filed a claim for "anticipatory nuisance" in the District Court for the Western District of Oklahoma. The following is an excerpt from the Court's decision.[39]

> The individual plaintiffs to this action[1] own land in Kingfisher and Canadian Counties, within the State of Oklahoma. Plaintiff Oklahoma Wind Action Association (OWAA) is an Oklahoma non-profit corporation. In this action, OWAA, purportedly on behalf of itself and its members, and the individual plaintiffs, bring tort claims for anticipatory nuisance against Defendant Kingfisher Wind LLC (Kingfisher) stemming from its construction and maintenance of a wind turbine farm in Kingfisher and Canadian Counties (hereinafter the Kingfisher Wind Project or KWP).

Before the Court is Kingfisher's Motion for Summary Judgment . . . on the Merits of Plaintiffs' Nuisance Claim. Plaintiffs have responded to both motions. The matter is fully briefed and at issue. The Court heard oral argument on the motions on September 27, 2016.

. . .

FACTUAL BACKGROUND

Kingfisher is in the business of constructing and operating industrial wind turbines. At issue in the present case is the Kingfisher Wind Project, which consists of 149 wind turbines. Plaintiffs contend the wind farm's construction and operation constitutes an anticipatory nuisance; specifically, Plaintiffs allege the wind farm causes and/or will cause adverse health effects, emits noise at a disconcerting level, destroys the natural landscape, interferes with the use and enjoyment of their property, and otherwise constitutes an annoyance due to "shadow flicker" and the casting of light. Plaintiffs seek permanent injunctive relief in the form of a 1.7 mile setback of the wind farm from any property owned by a plaintiff.

II. ANTICIPATORY NUISANCE

Oklahoma law provides a common law and statutory cause of action for private nuisance. By statute, "[n]uisance . . . consists in unlawfully doing an act, or omitting to perform a duty, which act or omission annoys, injures, or endangers the comfort, repose, health or safety of others; or, in any way renders other persons insecure in life, or in the use of property." Common law nuisance is defined more broadly as the "unwarrantable, unreasonable or unlawful use by a person of his own property to the injury of another." Under either a common law or statutory nuisance claim, a court may award, among other things, injunctive relief against a defendant who created, maintained, or owns a nuisance.

Oklahoma law further recognizes a claim for "anticipatory nuisance," i.e., a plaintiff can seek injunctive relief without having to wait for the actual infliction of a loss. However, the harm suffered must be irreparable—not compensable in money damages—and the evidence must be "clear and convincing that there is a reasonable probability of injury, not just a mere apprehension."

Generally, the conduct of a lawful business is not a nuisance per se. However, a lawful business may become a nuisance in fact if the place and manner of operation result in such substantial injury as to amount to an invasion of a homeowner's rights to repose and the enjoyment of their property.

The Court must consider whether Kingfisher has identified a lack of evidence on an essential element of Plaintiffs' anticipatory nuisance claim, and whether Plaintiffs have come forward with sufficient evidence to present a triable issue. If Plaintiffs' evidence is insufficient, it cannot, as a matter of law, achieve actual success on the merits, i.e., that the KWP is, or will constitute a nuisance. Further, to be entitled to its requested relief—a permanent injunction—Plaintiffs must not only show actual success on the merits, but must also make out a clear and unequivocal showing of the remaining elements of the equitable remedy.

For their anticipatory nuisance claim, Plaintiffs rely almost entirely on asserted adverse health effects and annoyance caused by the turbines. Kingfisher asserts that Plaintiffs have failed to present more than merely speculative proof of harm from the operation of the turbines.

On the record before it, the Court concludes that Plaintiffs have failed to make the required showing of likely harm. Specifically, Plaintiffs have failed to show a triable issue that there exists a reasonable probability an injury will occur as a result of the wind farm's operation. The harm alleged by Plaintiffs and their experts, at this juncture, is speculative at best, and in the Court's view, a reasonable trier of fact could not conclude, based on the evidence presented to date, that the shadow flicker or sound/infrasound from the turbines has caused or will cause adverse health effects to Plaintiffs. Moreover, the Court finds that the aesthetic concerns voiced by certain plaintiffs, based on the current record and absent any significant evidence of adverse health effects, are insufficient alone to constitute an actionable nuisance. At this late stage of the litigation, after a full opportunity to conduct discovery and marshal evidence, the injuries cited by Plaintiffs are simply too speculative to constitute harm sufficient under their anticipatory nuisance theory and, as discussed infra, to support the mandatory injunctive relief requested.[10]

For a party to obtain a permanent injunction, it must prove: (1) actual success on the merits; (2) irreparable harm unless the injunction is issued; (3) the threatened injury outweighs the harm that the injunction may cause the opposing party; and (4) the injunction, if issued, will not adversely affect the public interest. A permanent injunction is an extraordinary remedy, thus, the right to relief must be clear and unequivocal. As noted above, "a mere fear or apprehension of injury is insufficient." "Moreover, the Court recognizes that mandatory permanent injunctions are looked upon disfavorably and are generally only granted in compelling circumstances."

In addition to the failure of proof regarding the harm required to establish a claim for anticipatory nuisance, the balance of hardships as between the parties cannot reasonably be viewed as tipping in favor of Plaintiffs. Defendant correctly points out that, because Plaintiffs elected not to pursue preliminary injunctive relief, there was no legal impediment to its continued construction—and completion—of the wind farm during the pendency of this case. Thus, the KWP is now operational, at a cost of approximately $450,000,000. Plaintiffs' stated reason for eschewing the pursuit of preliminary injunctive relief early on in this case—that they would not likely have been able to afford an injunction bond—is unavailing (the Court has "wide discretion" in determining the amount of a bond). In the face of Plaintiffs' speculative evidence of injury from the KWP, the enormous cost and delay associated with the relief sought by Plaintiffs strongly militates against an injunction here, and compels the clear conclusion that the balance of hardships here does not tip in Plaintiffs' favor.

Questions

1. What is an "anticipatory nuisance" in Oklahoma? Did the judge allow the property owners to assert this?
2. Were the plaintiff landowners successful in alleging that they would suffer permanent injury? What did the judge say?
3. Did the Court balance the costs versus the benefits properly? Would it be possible to come to a different conclusion than the Court did here?
4. Common-law actions like this are rarely successful. What could the plaintiffs in this case have done differently to improve their chances against the windfarm?

NOTES

1. Emberson, L., He, K., Rockström, J., et al. 2012. Energy and Environment. In *Global Energy Assessment: Toward a Sustainable Future*, Chapter 3 Cambridge University Press, pp. 191–254.

2. See U.S. Geological Survey. *Which Country Has the Most Coal?* https://www.usgs.gov/faqs/which-country-has-most-coal.

3. See USEPA. *Deepwater Horizon – BP Gulf of Mexico Oil Spill.* https://www.epa.gov/enforcement/deepwater-horizon-bp-gulf-mexico-oil-spill.

4. For example, see the website for The United Nations Environment Programme. *How to Manage the Damage from Oil Spills.* https://www.unep.org/news-and-stories/story/how-manage-damage-oil-spills.

5. See the National Institute of Environmental Health Sciences. *Hydraulic Fracturing & Health.* https://www.niehs.nih.gov/health/topics/agents/fracking.

6. For example, see: Adgate, J.L., Goldstein, B.D., & McKenzie, L.M. 2014. Potential Public Health Hazards, Exposures and Health Effects From Unconventional Natural Gas Development. *Environmental Science & Technology, 48*(15), 8307–8320. https://doi.org/ 10.1021/es404621d.

7. For example, see Gross, M., & Mautz, R. 2014. *Renewable Energies*. Routledge; and Twidell, J. 2021. *Renewable Energy Resources*. Routledge.

8. See Maradin, D. 2021. Advantages and Disadvantages of Renewable Energy Sources Utilization. *International Journal of Energy Economics and Policy, 11*(3), 176–183.

9. See EnergySage. *The Advantages and Disadvantages of Renewable Energy.* https://www.energysage.com/about-clean-energy/advantages-and-disadvantages-of-renewable-energy/#advantages-of-renewable-energy.

10. See U.S. Energy Information Administration. *Solar Explained: Solar Energy and the Environment.* https://www.eia.gov/energyexplained/solar/solar-energy-and-the-environment.php.

11. Just Energy. *Solar Incentives and Rebates by State.* https://justenergy.com/solar/incentives-and-rebates-by-state/.

12. See McKenna, R., vd Leye, P. O., & Fichtner, W. 2016. Key Challenges and Prospects for Large Wind Turbines. *Renewable and Sustainable Energy Reviews, 53,* 1212–1221.

13. U.S. Energy Information Administration. *Wind Explained: Wind Energy and the Environment.* https://www.eia.gov/energyexplained/wind/wind-energy-and-the-environment.php.

14. U.S. Energy Information Administration. *Hydropower Explained.* https://www.eia.gov/energyexplained/hydropower/.

15. Zhang, A.T., & Gu, V.X. 2023. Global Dam Tracker: A Database of More Than 35,000 Dams with Location, Catchment, and Attribute Information. *Scientific Data, 111.* https://doi.org/10.1038/s41597-023-02008-2.

16. U.S. Geological Survey. *Three Gorges Dam, China is the World's Largest Hydro Facility.* https://www.usgs.gov/media/images/three-gorges-dam-china-worlds-largest-hydro-facility.

17. See the review in Akashie, E. 2022. *Advantages and Disadvantages of Hydroelectric Energy.* https://doi.org/10.13140/RG.2.2.31028.94081. https://www.researchgate.net/publication/359136049_Advantages_and_Disadvantages_of_hydroelectric_energy.

18. There is increasing evidence that hydroelectric dam construction can have profound environmental effects. See, for example: Schmutz, S., & Moog, O. (eds.). 2018. Dams: Ecological Impacts and Management. In *Riverine Ecosystem Management Science for Governing Towards a Sustainable Future*, Chapter 6, pp. 111–127; and Bauni, V., Schivo, F., Capmourteres, V., & Homberg, M. 2015. Ecosystem Loss Assessment Following Hydroelectric Dam Flooding: The Case of Yacyretá, Argentina. *Remote Sensing Applications: Society and Environment, 1,* 50–60.

19. See International Renewable Energy Agency. 2014. *Global Bioenergy Supply and Demand Projections.* A Working Paper for REmap 2030. www.irena.org.

20. U.S. Energy Information Administration. *Biomass Explained.* https://www.eia.gov/energyexplained/biomass/.

21. After U.S. Energy Information Administration. *Biomass Explained.* https://www.eia.gov/energyexplained/biomass/.

22. See U.S. Department of Energy. Energy.com. *DOE Releases Report Outlining How America Can Sustainably Produce More Than One Billion Tons of Biomass Per Year.* https://www.energy.gov/articles/doe-releases-report-outlining-how-america-can-sustainably-produce-more-one-billion-tons.

23. U.S. Department of Energy. Energy.com. *DOE Releases Report Outlining How America Can Sustainably Produce More Than One Billion Tons of Biomass Per Year.* https://www.energy.gov/articles/doe-releases-report-outlining-how-america-can-sustainably-produce-more-one-billion-tons.

24. UNEP. 2020. *Sustainable Bioenergy Use: A Clear Path to Biodiversity and Sustainable Economic Development.* https://www.unep.org/news-and-stories/story/sustainable-bioenergy-use-clear-path-biodiversity-and-sustainable-economic.

25. Kammen, D.M. 2004. *Renewable Energy, Taxonomic Overview. Encyclopedia of Energy.* Science Direct. Geothermal Energy. https://www.sciencedirect.com/topics/social-sciences/geothermal-energy.

26. There are several excellent reviews of the history of geothermal energy and its industrial uses. For example, see: Soltani, M., Moradi Kashkooli, F., Dehghani-Sanij, A. R., Nokhosteen, A., Ahmadi-Joughi, A., Gharali, K., . . . & Dusseault, M. B. 2019. A Comprehensive Review of Geothermal Energy Evolution and Development. *International Journal of Green Energy, 16*(13), 971–1009; and Sanner, B. 2016, September. Shallow Geothermal Energy–History, Development, Current Status, and Future Prospects. In *Proceedings of the European Geothermal Congress. T2*. Strasbourg, France. pp. 19–24.

27. U.S. Energy Information Administration. *Geothermal Explained: Use of Geothermal Energy*. https://www.eia.gov/energyexplained/geothermal/use-of-geothermal-energy.php.

28. EnergySage. *The Top Pros and Cons of Geothermal Energy*. https://www.energysage.com/about-clean-energy/geothermal/pros-cons-geothermal-energy/.

29. There are many sources available that review energy law and its history. A good starting point is the Legal Information Institute (LII) of Cornell Law School. See Wex.energy at https://www.law.cornell.edu/wex/energy.

30. 30 U.S.C. §181.

31. 16 U.S.C. §791-825r.

32. 42 U.S.C. Chapter 23.

33. 16 U.S.C. §2601 et seq.

34. Energy Policy Act of 2005, 42 USC §13201 et seq.

35. 42 U.S.C. §17001 et seq.

36. Amended 16 U.S.C. §2601 et seq., 42 U.S.C. §13201 et seq., and 42 U.S.C. §15801 et seq.

37. Public Law 117–58.

38. 136 Stat. 1366.

39. *Walker v. Kingfisher Wind*, LLC, 2016 WL 5947307 (2016).

Chapter 11

Military Actions and the Environment

Chapter Learning Outcomes
At the end of this chapter, the reader will be able to:

1. *Describe the historical use of petroleum-based weapons and changes to topography in combat.*
2. *Determine the ways that military bases can affect the environment.*
3. *List the environmental effects of military training activities.*
4. *Describe the environmental effects of contaminants, including hazardous and radioactive wastes, related to military combat.*
5. *Discuss the relationship between environmental laws and military actions.*

The relationship between the military and the environment is unique. Rarely do humans intentionally cause complete and lasting destruction of the environment as they do during times of war. In this chapter, we will explore some of the environmental impacts of military actions and warfare.[1] Then we will look at the modern military and how its actions affect the environment. Finally, we will discuss what happens when these issues confront the legal system.

11.1 THE HISTORY OF THE MILITARY AND THE ENVIRONMENT

The issues associated with warfare and the environment are as old as human civilization. History is replete with stories of burned forests, poisoned water supplies, or even landforms that were changed or destroyed, all in an attempt for one group to dominate another.

We don't know exactly when environmentally destructive warfare was used for the first time, but there are histories that date back to neolithic times that suggest that settlements and crops were burned by invading forces, causing immense human and environmental damage.

Petroleum-based mixtures are some of the most environmentally unfriendly materials used by militaries, both because of their destructive effects on ecosystems and because they remain in soils and water for centuries. These substances were used as accelerants for fire as early as the seventh century BCE, and probably even earlier. One of the first historically significant examples is the use of *Greek fire*, an early form of napalm, by the Byzantine people (eastern Romans who spoke Greek). The Byzantines used this seventh-century compound to repel Arab invasions for years. Greek fire was famously effective at sea where it had overwhelming effects on wooden ships (figure 11.1). Greek fire was eventually used in land-based warfare as well, with devastating effects on opponents and the environment (figure 11.2).[2]

In another type of early military action, landforms were altered for military purposes. An early example occurred when the Roman army attacked a Judean fortress built atop a flattened hill called Masada. Between the years 72 and 73 CE, the Romans built a 115-meter (380-foot) high ramp of natural

Figure 11.1 Illustrations of the Use of Greek Fire. Twelfth Century Image from an illuminated manuscript, the Madrid Skylitzes, showing Greek fire in use against the fleet of Thomas the Slav. The Greek caption above the left ship reads in English "the fleet of the Romans setting ablaze the fleet of the enemies." Public domain image. *Source*: Codex Skylitzes Matritensis, Bibliteca Nacional de Madrid, Vitr. 26–2, Bild-Nr. 77, f 34 v. b. (taken from Pászthory, p. 31). Unknown author

Figure 11.2 Greek Fire on Land as a "Hand-siphon." Use of a hand-siphon, a portable flame-thrower, from a siege tower equipped with a boarding bridge against the defenders on the walls (ninth or eleventh century). *Source*: The image is from a Medieval illumination in the manuscript Codex Vaticanus Graecus in 1605. Public domain image. Codex Vaticanus Graecus 1605 (taken from Pászthory, p. 31). Unknown author

rock, stones, and soil that allowed them to breach the Judean defenses. Much of the massive ramp remains to the present day.[3]

Other military landform alterations include redirecting water flow to flood enemy land. For example, in 1584, William of Orange ordered his Dutch rebels to open a series of gaps in the network of levies and dykes that kept the low-lying lands of the Netherlands free from flooding. The plan was to force Catholic troops off the land, but the effect was to allow salt water to enter low-lying Dutch land, destroying land and lives, and making it uninhabitable for centuries.

11.2 THE MILITARY DURING PEACETIME

Fortunately, most modern militaries spend the bulk of their time involved in activities other than combat. But even in peacetime, the military has a massive environmental footprint. For purposes of discussion, let's put these activities into two categories: military bases and training.

Military Bases

The environmental impacts of the construction of a military base depend on the nature of the site. The development of naval ports and shipyards is more likely to have a greater contamination risk to adjacent water bodies than the development of a land-based facility, which can be situated miles from water sources and surrounded by a natural vegetation buffer zone.[4]

In general, there are several types of environmental impacts associated with the construction of any complex infrastructure project, including military facilities. These impacts include habitat degradation, soil erosion, and chemical contamination. Site development usually requires the clearing of vegetation and trees, followed by intensive soil excavation and compaction, which alter natural landscapes and cause extensive damage to natural ecosystems. Furthermore, the removal of vegetation and soil excavation increases the potential for soil erosion and reduces water infiltration rates, altering landscape ecology by changing soil structure and chemistry, and increasing water runoff.

Chemical contamination of surface and groundwater often occurs from increased water runoff carrying sediments and chemicals associated with waste dumping and accidental chemical spills. Pollutants such as heavy metals, petroleum products, and a variety of hazardous and toxic chemical pollutants can be deadly to animal and plant populations and can alter natural community structures.

However, military installations, especially training bases, often have beneficial impacts on biodiversity as well. In particular, combat training facilities are often large and encompass a variety of environments and climates. Moreover, public and commercial access are usually restricted because of safety and security issues, which creates large tracts of land largely devoid of human contact, preserving them as wilderness areas. It has been estimated that over 200 federally listed endangered species inhabit U.S. military training areas.[5]

Military Training and the Environment

Training activities, especially live-fire exercises, and the use of vehicles (aircraft, ships, trucks, etc.) can lead to environmental damage that is comparable to actual combat. Military training facilities usually have live-fire shooting ranges, which lead to environmental degradation and contamination of the site. These sites are often associated with extensive heavy metal contamination, particularly with lead. Firing ranges for armor and artillery often experience contamination with heavy metals, nitrogen compounds, and even radioactive materials. These highly disturbed landscapes can suffer from degraded soil and are reduced to disturbance-tolerant plant and animal species. Of course, there are also issues associated with over-pressure injuries to wildlife from explosive detonations and live-fire operations.

Armored vehicles are extremely heavy, and their heavy weight causes extreme soil compaction, which can have detrimental impacts on soil and vegetation communities, especially in fragile environments like deserts (figure 11.3). Seasonal training can cause greater environmental damage, with higher rates of vegetation destruction. Training exercises in wooded areas can be particularly degrading on vegetation communities, as tracked vehicles are often used as bulldozers to clear paths and sightlines, and wildlife can be impacted directly.

Infantry training exercises can have a wide range of environmental impacts, depending on the size of the infantry and the nature of the exercise itself. Dismounted infantry can cause the destruction of vegetation, alter soil structure, and increase soil erosion. These exercises can negatively affect wildlife distribution in active training areas, where infantry presence can alter the behavior of animals, especially large mammal species, and alter ecosystems.

Military training areas are also used for fixed-wing aircraft and helicopter training. The largest environmental impact associated with aviation exercises is hitting and killing birds. Pollution of soil, surface water, and groundwater from munitions is similar to that in other military actions.

The main impacts of naval training exercises are caused by the generation of excessive noise pollution generated from sources such as mechanical and propeller noise, gun discharges, explosive detonations, and the use of sonar technologies. Sonar is known to impact the biology of large marine mammals, which rely on echolocation for most biological aspects of their lives,

Figure 11.3 Abrams Tank Firing Its Gun. An Army M1A2 Abrams tank participates in training at Fort Benning, Ga., Jan. 30, 2020. *Source*: U.S. Department of Defense. https://www.defense.gov/Multimedia/Photos/igphoto/2002243692/

and the use of sonar has been linked to disrupting their signaling abilities. Sonar has also been linked to mass stranding mortality events in whales, although the mechanism is unknown.

11.3 THE ENVIRONMENTAL EFFECTS OF COMBAT

It should come as no surprise that the greatest damage to the environment comes from actual military combat. Military combat is associated with the testing, production, transportation, and deployment of weapons. The potential for environmental destruction and contamination exists at each stage. Chemicals can be manufactured for use in weapons to cause direct human mortality or to alter landscapes to gain strategic tactical advantages that can expose the surrounding ecosystems to physical damage and exposure to potentially toxic chemicals.

Military activities also have the potential to contaminate the environment indirectly through various byproducts and spills associated with warfare, such as fuels and compounds used in vehicle maintenance. Various chemicals, including hydrocarbons and metals, can have immediate destructive and toxic effects that may also persist for long periods of time in soil, water, and the tissues of animals.[6]

How Has the Military Impacted the Environment?

It's beyond argument that the U.S. military has an extensive history of damaging the environment, both at home and abroad. Of course, this is true for most militaries in most countries. For our purposes, we will separate the impacts into four general categories: gas emissions (especially greenhouse gases), hazardous and toxic pollutants, radioactive materials, and direct combat-related impacts.

Gas Emissions

Gas emissions, especially greenhouse gases, resulting from military action have been documented in several studies.[7] Toxic gases have been released intentionally by warring parties in the past, such as the use of mustard gas in World War I, or the alleged use of sarin gas (military designation GB) during the Gulf War. But the release of other gases, like CO_2, VOCs, and methane, among many others, can cause both environmental and health issues, and are major contributors to climate change.

One study found that the U.S. military released roughly fifty-nine million metric tons of carbon dioxide and other greenhouse gases into the atmosphere during the year 2017.[8] If it were a country, this would place the U.S. military as the fifty-fifth largest contributor of greenhouse gas emissions in the world, putting it ahead of industrialized countries like Sweden and Portugal.

Energy consumption, and the associated release of greenhouse gases, is a major source of air pollution by the military. About 70 percent of all military energy consumption is for the use and movement of troops and weapons.

Taken as a whole, the Department of Defense is the world's largest institutional user of petroleum and, correspondingly, the single largest producer of greenhouse gases in the world.

Hazardous and Toxic Pollutants

The release of hazardous and toxic chemicals into the soil, water, and air seems to be an inevitable consequence of modern warfare. Military ordnance releases substances including heavy metals (lead, mercury, and barium) along with nitric oxides, perchlorates, and other potentially harmful substances. We discussed the impacts of these and related substances in chapter 9.

Pollution by toxic substances is a problem at many military installations and is a particular threat to military service members. There are ongoing investigations at several military bases like Camp Lejeune in North Carolina, where military personnel drank contaminated water for years, leading to many deaths and illnesses. A Newsweek report in 2014 found that the Pentagon's environmental programs were administering 39,000 contaminated sites around the country, including 141 Superfund sites that qualify for long-term cleanup grants from the federal government.[9]

An important source of hazardous and toxic waste release is the twenty-first century use of burn pits in Afghanistan, Iraq, and several other countries in the region. A burn pit is an open-air area used by the U.S. military outside of the United States to burn solid waste on a large scale. The types of waste burned can range from chemicals and human waste to food scraps and rubber. As a result, a variety of hazardous and toxic wastes are released into the air, water, and soil. There is every reason to believe that the long-term effects of burn pits on ecosystems are the same as the effects of other hazardous and toxic wastes as we discussed in chapter 9.

Toxic and hazardous materials in the form of herbicides have been used as defoliants since the twentieth century. Defoliants have also been used in warfare as a means to deprive an enemy of food crops and/or hiding cover, most notably by the United Kingdom during the Malayan Emergency and the United States in the Vietnam War. Defoliants were also used by Indonesian forces in various internal security operations. The most recognizable wartime defoliant was *agent orange*, a mixture of the herbicides 2,4,5-T and 2,4-D, which was used by the United States military against North Vietnamese forces who used the forests for cover. Defoliants have left an unfortunate legacy of injuries to people and destruction of ecosystems.

A long-term problem associated with herbicide exposure is bioaccumulation and the persistence of the chemicals in the environment. Over time, some herbicides break down into dioxin, a deadly chemical that can enter the bodies of wildlife. Herbicides and their byproducts can continue to cause environmental damage for decades.

Radioactive Materials

Radioactive substances are also a part of our military's legacy. Health and environmental damage from radioactive fallout resulting from nuclear weapons testing in the twentieth century remains a worldwide issue today. A 2002 CDC study found that any person who has lived in the United States since 1951 has been exposed to radioactive fallout from more than 500 nuclear weapons tests performed worldwide prior to the 1970s.[10]

The risks from radioactive wastes are especially acute in the western United States, where our nuclear weapons development and testing occurred, and where many uranium mines are found. These mines and debris sites are at risk for leaking due to time and environmental conditions. There are more than 160,000 abandoned uranium mines in the United States, many of which have polluted water supplies. In the Marshall Islands, an 18-inch concrete cap covering radioactive debris from U.S. nuclear testing is in danger of leaking into the sea due to rising sea levels.

An interesting issue involves the use of depleted uranium armor-piercing rounds in military actions in the Middle East. In truth, depleted uranium emits alpha particles, which lack sufficient energy to pass through the skin. As a result, exposure to the outside of the body is not considered a serious hazard. However, if depleted uranium is ingested or inhaled, it is a serious health hazard and is a potential environmental hazard.

Combat-Related Impacts

Certainly, the militaries of the world are responsible for direct impacts on the environment. As we mentioned, there are few human endeavors that have direct, intentional destructive effects on the environment like military combat. Weapons such as bombs, rockets, and artillery produce pressure waves that can have huge impacts on plants, wildlife, and entire ecosystems. Direct impacts from projectile weapons have caused tens of thousands of deaths of large mammals in Africa, Asia, and South America.

Examining the recent wars in Iraq, Afghanistan, and Pakistan, a recent study found that combat led to the degradation of multiple natural resources, including contamination of water resources in war zones by oil from military vehicles, and deforestation in Afghanistan as a result of illegal logging. Along with the degradation of natural resources in these countries and the radical destruction of forest cover, the animal and bird populations have also been adversely affected.[11] According to the United Nations, military activity has also contributed to extreme desertification in Iraq.

Modern military combat releases many hazardous, toxic, and even radioactive materials into the environment, as we discussed above. Ammunition and ordnance release metals, hazardous chemicals of several kinds, and even depleted uranium into air, soil, and water. Defoliants, other herbicides, and pesticides of various kinds contribute to changes and degradation of ecosystems.

There have been several attempts to identify and understand the environmental costs of military combat, and to look for ways to reduce or eliminate these costs. It is clear that the military often faces conflicting roles in balancing the needs of national defense against the protection of the environment. A 2008 study found that environmental considerations by the U.S. Army rarely received high priority because they competed with other war combat concerns. The study found that putting a low priority on the environmental consequences of actions creates missed opportunities for the military to make strategic use of environmental issues. Poor environmental conditions can also lead to further instability during post-conflict operations.[12]

Studies are currently underway in all branches of the U.S. military to find ways to minimize the environmental effects of military operations of all kinds, but these studies are mostly in their infancy, and their future is somewhat uncertain. On the other hand, the Pentagon has announced several actions to address climate change, which we will discuss in chapter 12.

11.4 ENVIRONMENTAL LAWS AND THE MILITARY

Strictly speaking, all environmental laws apply to the military through the Department of Defense in the same way that they apply to all federal agencies. In fact, all local, state, and federal laws apply to the military and military personnel. It's more complicated, though.

It can be argued that strict enforcement of environmental laws and regulations is an impediment to the military's training and combat abilities. From the beginning of major U.S. environmental legislation in the 1970s, there have been significant challenges to military compliance with the corresponding regulations associated with specific provisions of NEPA, the CAA, the CWA, and the Endangered Species Act.[13]

Specific provisions within each of these environmental laws provide exemptions from specific environmental regulations in order to assist the military. For example, a Clean Air Act provision permits the president to exempt any emission source if they determine it to be "in the paramount interest of the United States."[14] Most of these exemptions are granted for a period of two years with a reassessment at the end of that period for an additional year of exemption. The bottom line is that the president can assess any relevant security interest and grant an exemption. In addition, Council on Environmental Quality regulations relax NEPA requirements for emergency circumstances.

Under the Endangered Species Act, exemptions from species' protections are authorized by the opinion of the appropriate Secretary and by action of an Executive "Endangered Species Committee." That Committee must grant an exemption if the Secretary of Defense finds it necessary for reasons of national security.[15]

Under the Clean Water Act, the president may exempt any executive branch effluent source if she/he determines it to be "in the paramount interest of the United States," except that no exemption may be granted from the national standards for *new* sources and toxic and pretreatment effluent standards. An exemption may not exceed one year but may be extended for up to one year. In addition, the president may, if he/she determines it to be "in the paramount interest of the United States," exempt any weapons, equipment, aircraft, vessels, vehicles, or other classes or categories of property of the military that are uniquely military in nature.[16] Under the CWA's wetlands provision, the discharge of dredged or fill material into waters of the United States for certain military purposes does not require a Department of the Army permit.

Under the Safe Drinking Water Act §1447, the president may exempt any executive-branch facility, including military facilities, from SDWA requirements if he/she determines it to be "in the paramount interest of the United States."[17]

Executive branch exemptions also occur under many other federal environmental laws. Section 6001 of the Resource Conservation and Recovery Act (RCRA) allows the president to exempt any executive-branch solid waste management facility, including all military facilities, from RCRA's solid and hazardous waste requirements if she/he determines it to be "in the paramount interest of the United States."[18] Furthermore, §9008 of RCRA allows the president to exempt any executive-branch underground storage tank from RCRA requirements if he/she determines it to be "in the paramount interest of the United States."[19]

Many other federal environmental laws permit similar exemptions to military actions by the president. Under the CERCLA, the "Superfund law" §120(j) the president can issue orders regarding response actions at any specified CERCLA (Superfund) site or facility of the Department of Energy or the

Department of Defense "as may be necessary to protect the national security interests of the United States," including the military, from emission controls if he/she determines it to be "in the paramount interest of the United States."[20] Under the federal Noise Control Act §4 the president may exempt any single activity or facility or noise emission source of an executive agency, including the military, from emission controls if she/he determines it to be "in the paramount interest of the United States."[21]

In each of these cases, the Department of Defense, through the president, is provided significant statutory exemptions from compliance with federal environmental laws. These exemptions were created by Congress to allow the president and the military sufficient lack of restrictions in matters of national security.

Other federal laws are designed to provide support for individuals who were harmed by military actions on the environment. A recent example is the Sergeant First Class Heath Robinson Honoring our Promise to Address Comprehensive Toxics Act of 2022, known as the Honoring our PACT Act of 2022, or PACT.[22] The PACT law expands and extends eligibility for VA health care for veterans with toxic exposures and veterans of the Vietnam, Gulf War, and post-9/11 eras, with a focus on exposure by agent orange and burn pit exposures, among others.

* * *

Review Questions

1. Discuss the historical use of petroleum-based weapons and changes in topography in military actions.
2. What are the major impacts of military bases on the environment?
3. Describe the effects of military training activities on the environment.
4. Discuss the short-term and long-term effects of military wastes, including hazardous, toxic, and radioactive materials.
5. Explain the relationship between major environmental laws and military actions.

Case Analysis

We saw in chapter 1 that the National Environmental Policy Act requires the preparation of an impact statement before many federal actions that might affect the environment. We saw in chapter 4 that statutes like the Endangered Species Act and the Marine Mammal Protection Act create strong presumptions against taking a protected species. But how do these laws apply to actions by the military, especially when there is no direct conflict involved?

In 2008, the U.S. Supreme Court answered these questions under one set of circumstances. At issue was the U.S. Navy's use of active sonar in large-scale training activities in the Pacific Ocean off the coast of California. It is known that active sonar can interfere with the activities of endangered marine mammals. The case is complicated because the Navy had agreed to a series of actions to protect the mammals, but the plaintiffs (several environmental groups including the Natural Resources Defense Council, or NRDC) argued that the actions were not sufficient. The Circuit Court agreed with the plaintiffs, and the 9th Circuit Court of Appeals agreed, in principle. The Navy sued, leading to the U.S. Supreme Court decision in *Winter v. NRDC*.[23] Extracts of that decision, written by Chief Justice Roberts, are below.

> Antisubmarine warfare is currently the Pacific Fleet's top war-fighting priority. Modern diesel-electric submarines pose a significant threat to Navy vessels because they can operate almost silently, making them extremely difficult to detect and track. Potential adversaries of the United States possess at least 300 of these submarines.
>
> The most effective technology for identifying submerged diesel-electric submarines within their torpedo range is active sonar, which involves emitting pulses of sound underwater and then receiving the acoustic waves that echo off the target. Active sonar is a particularly useful tool because it provides both the bearing and the distance of target submarines; it is also sensitive enough to allow the Navy to track enemy submarines that are quieter than the surrounding marine environment. This case concerns the Navy's use of "mid-frequency active" (MFA) sonar, which transmits sound waves at frequencies between 1 kHz and 10 kHz.
>
> . . .
>
> The waters off the coast of southern California (SOCAL) are an ideal location for conducting integrated training exercises, as this is the only area on the west coast that is relatively close to land, air, and sea bases, as well as amphibious landing areas. At issue in this case are the Composite Training Unit Exercises and the Joint Tactical Force Exercises, in which individual naval units (ships, submarines, and aircraft) train together as members of a strike group.
>
> Sharing the waters in the SOCAL operating area are at least 37 species of marine mammals, including dolphins, whales, and sea lions. The parties strongly dispute the extent to which the Navy's training activities will harm those animals or disrupt their behavioral patterns. The Navy emphasizes that it has used MFA sonar during training exercises in SOCAL for 40 years, without a single documented sonar-related injury to any marine mammal. The Navy asserts that, at most, MFA sonar may cause temporary hearing loss or brief disruptions of marine mammals' behavioral patterns.
>
> The plaintiffs . . . contend that MFA sonar can cause much more serious injuries to marine mammals than the Navy acknowledges, including permanent hearing loss, decompression sickness, and major behavioral disruptions.

According to the plaintiffs, several mass strandings of marine mammals (outside of SOCAL) have been "associated" with the use of active sonar. They argue that certain species of marine mammals—such as beaked whales—are uniquely susceptible to injury from active sonar; these injuries would not necessarily be detected by the Navy, given that beaked whales are "very deep divers" that spend little time at the surface.

. . .

The procedural history of this case is rather complicated. The Marine Mammal Protection Act of 1972 (MMPA) generally prohibits any individual from "taking" a marine mammal, defined as harassing, hunting, capturing, or killing it. 16 U. S. C. §§1362(13), 1372(a). The Secretary of Defense may "exempt any action or category of actions" from the MMPA if such actions are "necessary for national defense." §1371(f)(1). In January 2007, the Deputy Secretary of Defense—acting for the Secretary—granted the Navy a 2-year exemption from the MMPA for the training exercises at issue in this case. The exemption was conditioned on the Navy adopting several mitigation procedures, including: (1) training lookouts and officers to watch for marine mammals; (2) requiring at least five lookouts with binoculars on each vessel to watch for anomalies on the water surface (including marine mammals); (3) requiring aircraft and sonar operators to report detected marine mammals in the vicinity of the training exercises; (4) requiring reduction of active sonar transmission levels by 6 dB if a marine mammal is detected within 1,000 yards of the bow of the vessel, or by 10 dB if detected within 500 yards; (5) requiring complete shutdown of active sonar transmission if a marine mammal is detected within 200 yards of the vessel; (6) requiring active sonar to be operated at the "lowest practicable level"; and (7) adopting coordination and reporting procedures.

In February 2007, the Navy issued an EA [Environmental Assessment under NEPA] concluding that the 14 SOCAL training exercises scheduled through January 2009 would not have a significant impact on the environment.

' ' '

Shortly after the Navy released its EA, the plaintiffs sued the Navy, seeking declaratory and injunctive relief on the grounds that the Navy's SOCAL training exercises violated NEPA, the Endangered Species Act of 1973 (ESA), and the Coastal Zone Management Act of 1972 (CZMA). The District Court granted plaintiffs' motion for a preliminary injunction and prohibited the Navy from using MFA sonar during its remaining training exercises.

. . .

The Navy filed an emergency appeal, and the Ninth Circuit stayed the injunction pending appeal. 502 F. 3d 859, 865 (2007)

. . .

The Navy then sought relief from the Executive Branch.. . . . The Council on Environmental Quality (CEQ) authorized the Navy to implement "alternative arrangements" to NEPA compliance in light of "emergency circumstances." The CEQ determined that alternative arrangements were appropriate because the District Court's injunction "create[s] a significant and unreasonable risk that

Strike Groups will not be able to train and be certified as fully mission capable." Under the alternative arrangements, the Navy would be permitted to conduct its training exercises under the mitigation procedures adopted in conjunction with the exemption from the MMPA. The CEQ also imposed additional notice, research, and reporting requirements.

. . .

Captain Martin May—the Third Fleet's Assistant Chief of Staff for Training and Readiness—emphasized that the use of MFA sonar is "mission-critical." . . . We accept these officers' assertions that the use of MFA sonar under realistic conditions during training exercises is of the utmost importance to the Navy and the Nation.

. . .

These interests must be weighed against the possible harm to the ecological, scientific, and recreational interests that are legitimately before this Court. Plaintiffs have submitted declarations asserting that they take whale watching trips, observe marine mammals underwater, conduct scientific research on marine mammals, and photograph these animals in their natural habitats. Plaintiffs contend that the Navy's use of MFA sonar will injure marine mammals or alter their behavioral patterns, impairing plaintiffs' ability to study and observe the animals.

While we do not question the seriousness of these interests, we conclude that the balance of equities and consideration of the overall public interest in this case tip strongly in favor of the Navy. For the plaintiffs, the most serious possible injury would be harm to an unknown number of the marine mammals that they study and observe. In contrast, forcing the Navy to deploy an inadequately trained antisubmarine force jeopardizes the safety of the fleet. Active sonar is the only reliable technology for detecting and tracking enemy diesel-electric submarines, and the President—the Commander in Chief—has determined that training with active sonar is "essential to national security."

The public interest in conducting training exercises with active sonar under realistic conditions plainly outweighs the interests advanced by the plaintiffs. Of course, military interests do not always trump other considerations, and we have not held that they do. In this case, however, the proper determination of where the public interest lies does not strike us as a close question.

. . .

President Theodore Roosevelt explained that "the only way in which a navy can ever be made efficient is by practice at sea, under all the conditions which would have to be met if war existed." We do not discount the importance of plaintiffs' ecological, scientific, and recreational interests in marine mammals. Those interests, however, are plainly outweighed by the Navy's need to conduct realistic training exercises to ensure that it is able to neutralize the threat posed by enemy submarines. The District Court abused its discretion by imposing a 2,200-yard shutdown zone and by requiring the Navy to power down its MFA sonar during significant surface ducting conditions. The judgment of the Court of Appeals is reversed, and the preliminary injunction is vacated to the extent it has been challenged by the Navy.

Questions

1. The Navy repeatedly argued that it had taken precautions and that any damage to marine mammals would be minor and temporary. What is the basis of their argument?
2. The plaintiff environmental groups argue that the damage to marine mammals is more severe and permanent in some cases. What evidence supports their argument? Who is right, the Navy or the NRDC?
3. Since the Secretary of the Navy granted a 2-year exemption from the MMPA restrictions, and the CEQ said the restrictions would allow the Navy to proceed, what did the NRDC expect to accomplish?
4. The Supreme Court used a balancing test of the requirements of the military versus the effects on the environment. What law(s), if any, allow it to do that?
5. Some commentators have argued that the Supreme Court is "legislating from the bench" in this case. In other words, the Court is substituting its judgment for a law that does not exist. Are the commentators right? If the commentators are right, does that make the Supreme Court's decision wrong in this case?

NOTES

1. This discussion is not intended to be comprehensive. If you are interested in pursuing these issues, there are several excellent sources of information: Wilcox, W.A. 2007. *The Modern Military and the Environment.* Government Institutes; Dycus, W. 1996. *National Defense and the Environment.* University Press of New England; and Closmann, C.E. (Ed.). 2009. *War and the Environment: Military Destruction in the Modern Age.* Texas A&M University Press.

2. An excellent discussion of Greek fire and other ancient warfare techniques is in Mayor, A. 2022. *Greek Fire, Poison Arrows, and Scorpion Bombs: Unconventional Warfare in the Ancient World.* Princeton University Press.

3. See Magness, J. 2019. *Masada: From Jewish Revolt to Modern Myth.* Princeton University Press.

4. An exhaustive review is in Lawrence, M.J., Stemberger, H.L.J., Zolderdo, A.J., Struthers, D.P., & Cooke, S.J. 2015. The Effects of Modern War and Military Activities on Biodiversity and the Environment. *Environmental Reviews, 23*(4), 443–460. https://doi.org/10.1139/er-2015-0039.

5. There are more endangered species per area within military installations than any other federally managed lands in the U.S. See Zentelis, R., & Lindenmayer, D. 2014. Bombing for Biodiversity—Enhancing Conservation Values of Military Training Areas. *Conservation Letters, 8*(4), 299–305. https://doi.org/10.1111/conl.12155.

6. See Machlis, G.E., & Hanson, T. 2008. Warfare Ecology. *BioScience, 58,* 729–736. https://doi.org/10.1641/B580809.

7. For example, see Light, S.E. 2014. The Military-Environmental Complex. *Boston College Law Review, 55,* 879–946.

8. Crawford, N. 2019. *Costs of War*. Watson Institute for International and Public Affairs, Brown University. https://watson.brown.edu/costsofwar/.

9. Nazaryan, A. 2014. The U.S. Department of Defense Is One of the World's Biggest Polluters. *Newsweek Magazine*, July 17. https://www.newsweek.com/2014/07/25/us-department-defence-one-worlds-biggest-polluters-259456.html.

10. National Cancer Institute and the Center for Disease Control. 2002 (updated in 2014). *Radioactive Fallout from Global Weapons Testing*. https://www.cdc.gov/nceh/radiation/fallout/rf-gwt_home.htm.

11. Crawford, N. 2019. *Costs of War*. Watson Institute for International and Public Affairs, Brown University. https://watson.brown.edu/costsofwar/.

12. Mosher, D.E., Lachman, B.E., Greenberg, M.D., Nichols, T., Rosen, B., & Willis, H.H. 2008. *The Army's Green Warriors: Environmental Considerations in Contingency Operations*. Rand Corporation, Arroyo Center.

13. Nevitt, M.P. 2015. Environmental Law in Military Operations. In Corn, G.S., VanLandingham, R.E., & Reeves, S.R. (eds.) *U.S. Military Operations: Law, Policy, and Practice*. Oxford Academic; online ed. (17 December 2015).

14. Clean Air Act, 42 USC § 7418(b).

15. 16 U.S.C. § 1536 (e).

16. 33 U.S.C. § 1323.

17. 42 U.S.C. § 300j-6.

18. 42 U.S.C. § 6961.

19. 42 U.S.C. § 6991f).

20. 42 U.S.C. § 9620(j).

21. 42 U.S.C. § 4903.

22. Pub. L. 117–168, Statutes at Large 136 Stat. 1759.

23. *Winter v. Natural Resources Defense Council*, Inc., 555 U.S. 7, 129 S. Ct. 365 (2008).

Chapter 12

Global Climate Change

Chapter Learning Outcomes

At the end of this chapter, the reader will be able to:

- *Describe what is meant by climate change.*
- *Summarize the role of greenhouse gases in climate change.*
- *Predict changes in temperature and the hydrologic cycle due to climate change.*
- *Explain the rapid rise of temperatures in the Arctic.*
- *List the effects of climate change on ecosystems and agriculture.*
- *Describe the major U.S. laws and regulations that regulate climate change.*
- *Illustrate the role of international agreements on climate change.*
- *Identify ways to slow global climate change.*

Our final chapter will be an opportunity to review many of the principles that we discussed in previous chapters. Specifically, we will explore the phenomenon of global climate change as it is affecting all of us. We will begin by discussing some of the scientific evidence for global climate change in both long and short terms and examine some of the consequences of climate change as they impact humans and ecosystems. Then we will look at the national and international framework of laws that regulate our approaches to climate change.

To begin, what exactly do we mean by "climate change"? We discussed the components of climate in chapter 6. We can think of climate change as a long-term change in the average weather patterns that have come to define earth's local, regional, and global climates. These changes have a broad range of observed effects on ecosystems and on human health.

Here is an example. Scientists have determined that global temperatures rose about 1.98°F (1.1°C) from 1901 to 2022,[1] but climate change refers to more than an increase in temperature. It also includes changes in weather patterns like drought and flooding, increases in storm intensity, sea level rise, and much more. Important human requirements such as healthcare, agriculture, and transportation, as well as wildlife and ecosystems, are all experiencing the dramatic effects of a changing climate.[2]

We know that climates on earth have always changed, but they are now changing at a pace far faster than any changes that we have seen in the past. What is causing these rapid changes? Certainly, some change is the result of natural events, such as volcanic activity. But it is very likely that most of the climate change in the current era is the result of human activities, such as increased concentrations of greenhouse gases in the atmosphere. We will discuss these and other factors in this chapter.

12.1 THE CAUSES OF CLIMATE CHANGE

We know that the climate of earth has always changed over time. We base this on scientific evidence from a variety of sources. For example, fossils from the Cretaceous period (144 to 65 million years ago) show that earth was much warmer than it is today. Fossilized plants and animals that normally live in warm environments were found at much higher latitudes at that time than would be possible today. For instance, breadfruit trees (*Artocarpus altilis*), now found on tropical islands, grew as far north as Greenland. Let's explore some of the factors that cause climate changes both in the short and long terms.

Greenhouse Gases and Climate Change

We explored the greenhouse effect and greenhouse gases in chapter 6. Recall that radiant energy from the sun in the form of ultraviolet radiation is trapped at the earth's surface by a layer of *greenhouse gases* in the atmosphere. The greater the concentration of these gases, the greater the increase in temperature.

We learned in chapter 6 that the major greenhouse gases are carbon dioxide (CO_2), methane, ozone (O_3), nitrous oxide (N_2O), chlorofluorocarbons (CFCs), and water vapor. All of these greenhouse gases are important, but it is CO_2 levels that are most significant because atmospheric levels have increased dramatically in recent decades. These increases have driven much of the climate change the earth is experiencing. The increases are primarily the result of burning fossil fuels by humans, although some changes are attributable to natural events like forest fires and volcanic activity. Figure 12.1 shows atmospheric CO_2 levels for about 10,000 years. Note the dramatic increase in CO_2 levels in the last century, corresponding to increases in fossil fuel use by humans for transportation and industry.

The CO_2 graph in figure 12.1 is convincing, but how do we get the data? The answer is that paleoclimatologists extract the data from ice cores from

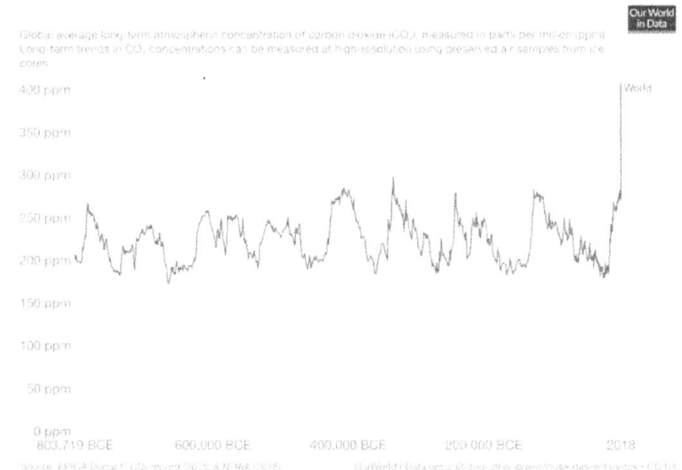

Figure 12.1 Graph Showing CO_2 Concentration (Parts per Million) in the Global Atmosphere over Time. *Source*: Wikimedia Commons. Our World in Data. https://creativecommons.org/licenses/by/4.0/

Figure 12.2 An Ice Core from West Antarctica Contains a Record of Climate for over 20,000 years. NASA. Core questions: An introduction to ice cores. *Source*: https://science.nasa.gov/science-research/earth-science/climate-science/core-questions-an-introduction-to-ice-cores/

ice sheets (figure 12.2). Ice sheets and glaciers near earth's North and South Poles formed from years and years of accumulating snowfall. The weight of each year's snowfall compresses the previous layers of snow, and after many years, the pressure helps to form glacial ice. In some areas, these layers result in ice sheets that are several miles (several kilometers) thick. Within these deep layers of ice are tiny bubbles that preserve the air in the climate at the time the ice forms. By analyzing the gases trapped in the bubbles in each ice layer, scientists can measure the exact gas content from a specific period of time. The dark band in the ice core in figure 12.2 is from West Antarctica, which shows a layer of volcanic ash that settled on the ice sheet approximately 21,000 years ago.

12.2 THE EFFECTS OF CLIMATE CHANGE

The specific consequences of climate change take several forms. We will investigate temperature changes ("global warming"), changes in the global hydrological cycle and weather, melting glaciers and sea level rise, ocean acidification, and effects on organisms and ecosystems.

Temperature Change

Earth has experienced at least four ice ages in the past 500,000 years. During these periods, earth's temperature decreased, causing an expansion of ice sheets and glaciers, and a decrease in sea levels. These ice ages alternated with periods of relative warmth lasting thousands of years, when glaciers retreated and sea levels rose. The most recent Ice age began about two million years ago and peaked about 20,000 years ago. The ice caps began retreating again about 18,000 years ago, and this continues to the present day.

However, these temperature changes have all been very gradual, measured over hundreds and thousands of years. The concern today is that temperatures are increasing extremely rapidly. Figure 12.3 illustrates temperature changes over the past 2,000 years. Note the dramatic increase in global temperatures beginning in the twentieth century.

If we examine recent temperature changes more closely, it is apparent that global temperature changes have not been uniform across the globe. Figure 12.4 shows surface air changes from 1973 to 2023, and it is clear that temperatures in the far North are the most extreme, exceeding 7°F (4°C) in areas near the Arctic Circle. By contrast, there are areas in the far South, especially some areas near the Antarctic, that have actually experienced lower temperatures over the same time period.

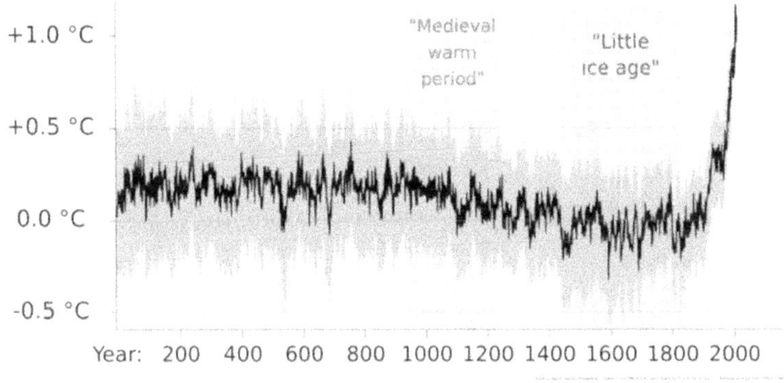

Figure 12.3 Global Average Temperature Changes over the Past 2,000 Years. Note the dramatic increase in the twentieth century. *Source*: Wikimedia Commons. RCraig09. Own work, CC BY-SA 4.0, https://commons.wikimedia.org/w/index.php?curid=87832845

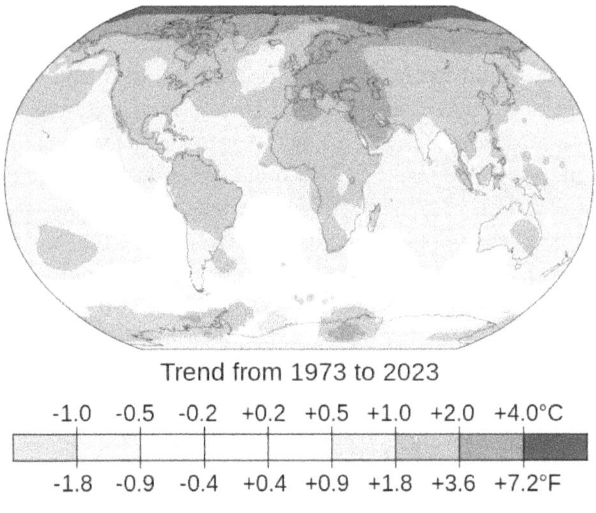

Figure 12.4 Surface Air Temperature Change over the Past 50 years. *Source*: Wikimedia Commons. NASA. https://data.giss.nasa.gov/gistemp/maps/index_v4.html

The reason temperature increases are most extreme in the far North has an interesting explanation. We know that ice reflects sunlight, while water absorbs it, heating the water. As more sea ice disappears, the underlying ocean surface is exposed. The much darker ocean surface absorbs sunlight instead of reflecting it, allowing much more heat to enter the Arctic system, and melting more ice. This is an example of a *positive feedback system* (or "vicious cycle") that increases temperatures in the North much faster than is experienced anywhere else on the planet. As we will discover shortly, this melting sea ice has profound effects on animals in these important ecosystems.

Figure 12.5 Map of Regions of the Globe where Natural Disasters Will Occur due to Climate Change. *Source*: Wikimedia Commons. KVDP. Own work. https://creativecommons.org/licenses/by-sa/3.0/deed.en

Changes in the Hydrologic Cycle and Weather

We discussed the water cycle, or hydrologic cycle, in chapter 3. Recall that the water cycle stores and distributes the earth's fixed water supply. Because there is a large reservoir of water in the atmosphere as well as in the hydrosphere, especially the oceans, the hydrologic cycle has a major impact on the earth's weather and climate.

There have been conspicuous geographic changes in total annual precipitation around the planet, with some areas experiencing severe and long-term drought and others experiencing increased annual precipitation. The frequency and intensity of storms increase as the atmosphere warms and is able to hold more water vapor. Parts of the world can expect an increase in both the frequency and intensity of storms, with increases in hurricanes in some areas, flooding in others, and droughts in still others. Figure 12.5 shows the world map with areas of increased extreme weather events.

We are already seeing increases in extreme weather.[3] In the year 2023, the United States alone experienced twenty-eight separate weather and climate disasters costing at least one billion dollars. That number put 2023 into first place for the highest number of billion-dollar disasters in a calendar year. The disasters in 2023 included:[4]

- 1 winter storm/cold wave event (across the northeast U.S. in early February).
- 1 wildfire event (firestorm destroying the town of Lahaina on Maui Island of Hawaii).
- 1 drought and heat wave event (focused across the central and southern United States).

- 4 flooding events (in California, Florida, and across the eastern and northeastern United States).
- 2 tornado outbreaks (across the central and eastern United States).
- 2 tropical cyclones (Idalia in Florida and Typhoon Mawar in Guam).
- 17 severe weather/hail events (across many parts of the country).

A 2024 report from the United Nations Convention to Combat Desertification reported that drylands, or those areas in which 90 percent of moisture is lost to evaporation, now account for 40 percent of the land on earth, excluding Antarctica.[5] The report indicates that 77.6 percent of land became permanently drier between 1990 and 2020. As an area's climate becomes drier, its ability to return to previous wetter conditions is lost. As a result, the drier climates now affecting lands across the globe will not return to previous, wetter conditions, and this change affects ecosystems worldwide.

All evidence indicates that extreme weather events will continue to increase, especially if global temperatures continue to climb.

Melting Glaciers and Sea Level Rise

Melting glaciers and sea level rise are directly related. There is ample evidence that glaciers are retreating. Figure 12.6 shows massive losses of glacial mass worldwide in "reference" glaciers.

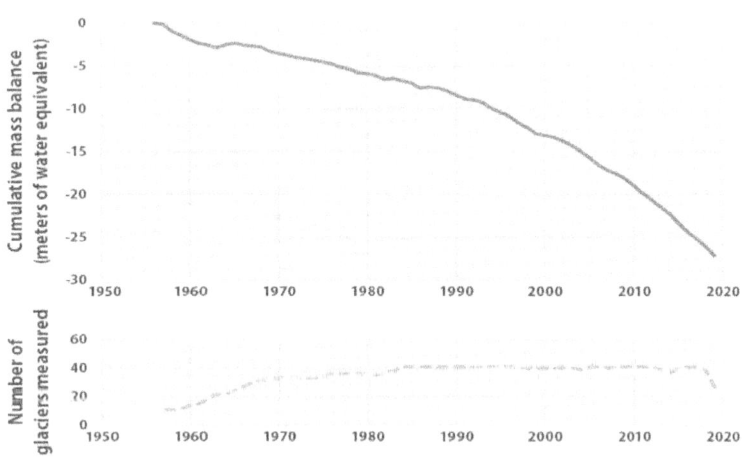

Figure 12.6 Average Cumulative Mass Balance of "Reference" Glaciers Worldwide, 1956–2019. *Source*: USEPA. Climate Change Indicators: Glaciers. https://www.epa.gov/climate-indicators/climate-change-indicators-glaciers

The bottom graph in figure 12.6 shows the number of glaciers measured. The sharp decline in 2019 resulted from the complete loss of several reference glaciers.

The loss of glaciers is directly related to increases in global temperatures. But melting glaciers lead to another issue. Glacial meltwater flows into the oceans of the world. This has two important effects on the oceans.[6]

First, ocean currents are affected by the decrease in ocean salinity that results when freshwater glaciers melt. Due to the difference in density between freshwater and saltwater, the influx of lower-density glacial melt may actually decrease the flow rate of the Atlantic Ocean's Gulf Stream. This could have a devastating effect on parts of Europe that rely on a strong Gulf Stream to bring warm water (and warm weather) to Europe. With a decreased Gulf Stream, temperature and rainfall could be affected, and agriculture impaired. Similar changes in currents can be anticipated in all of the earth's major oceans.

The second effect of melting glaciers is an increase in sea levels worldwide. Coastal flooding due to rising sea levels has already become a major global issue. It leads to the inundation of coastal communities, widespread beach erosion, and contamination of drinking water supplies.

There is ample evidence that sea levels have risen by an average of over eight inches since the late 1800s (Figure 12.7). Rising sea levels put coastal communities at great risk and has led to mass relocations of citizens from coastal areas in several parts of the world. In the United States, several models project that average sea level rise could be 2.2 meters (7.2 feet) by 2100 and 3.9 meters (13 feet) by 2150.[7]

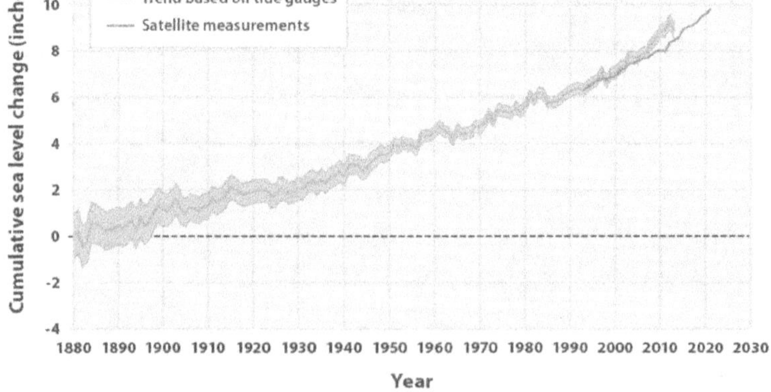

Figure 12.7 Global Average Absolute Sea Level Change (in inches), 1880–2021. *Source*: USEPA. Climate Change Indicators: Sea Level. https://www.epa.gov/climate-indicators/climate-change-indicators-sea-level

The effects of sea level rise can be devastating to coastal ecosystems. For example, many coastal freshwater wetlands have been inundated by saltwater, making the wetlands unlivable for many species. Rising sea levels can cause destructive erosion, aquifer contamination, and lost habitat for fish, birds, and plants.

Ocean Acidification

A lesser-known but equally important effect of climate change is the acidification of the world's oceans.[8] As the concentration of carbon dioxide in the atmosphere has increased over the past two centuries, the pH of surface ocean waters has fallen by 0.1 pH units.[9] This increase is far worse than it appears. Because the pH scale is logarithmic, a change of pH of 0.1 represents approximately a 30 percent increase in acidity. Figure 12.8 shows changes in pH levels measured in Hawaii, where the pH decreased by about 0.07 in just thirty-four years from 1988 to 2022.

The oceans of the world absorb about 30 percent of the carbon dioxide that is released into the atmosphere. As levels of atmospheric CO_2 increase, the amount of CO_2 absorbed by the oceans also increases. When CO_2 is absorbed by seawater, a series of chemical reactions take place, resulting in an increase in the concentration of hydrogen ions (H^+), which makes the water more acidic. This process has far-reaching implications for the ocean and the creatures that live there.

The chemistry of ocean acidification works like this: First, CO_2 from the atmosphere reacts with water in the oceans to produce carbonic acid (H_2CO_3).

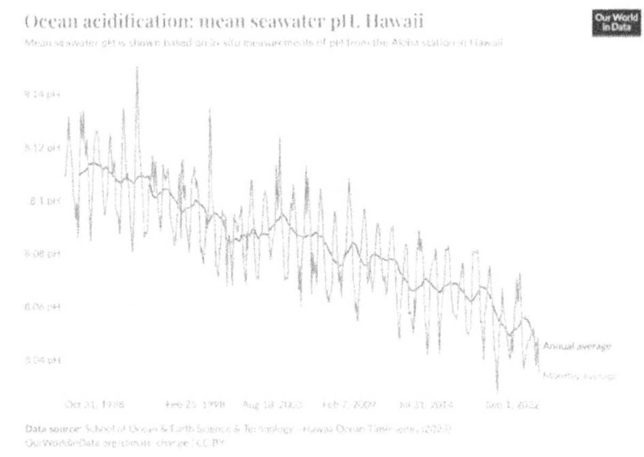

Figure 12.8 Changes in Ocean pH in Hawaii from 1988 to 2022. *Source*: Wikimedia Commons. Our World in Data

The carbonic acid in the ocean then dissociates into hydrogen ions (H^+) and bicarbonate ions (HCO_3^-). It is the presence of the hydrogen ions that makes the oceans more acidic. The chemical processes are illustrated and explained in figure 12.9.

The effects of ocean acidification on ocean wildlife are profound. In some cases, the behavior of fishes is changed by acidification. For example, research shows that juvenile clownfish leave the protection of corals and are more vulnerable to predators when ocean pH decreases.[10]

Many invertebrate species have shells or skeletons composed largely of calcium carbonate ($CaCO_3$). This calcium carbonate reacts with H^+ ions in acidified oceans to either break down calcium carbonate or prevent its formation. We can see this change in vulnerable species like corals, where ocean acidification along with ocean warming is causing massive death and bleaching of corals (figure 12.10).[11]

Another invertebrate affected by calcium carbonate loss is the pteropod, or sea butterfly, a tiny sea snail. Pteropods are an important part of many food chains. Researchers placed pteropod shells in seawater with pH and carbonate levels projected for the year 2100, the shells slowly dissolved after 45 days.[12]

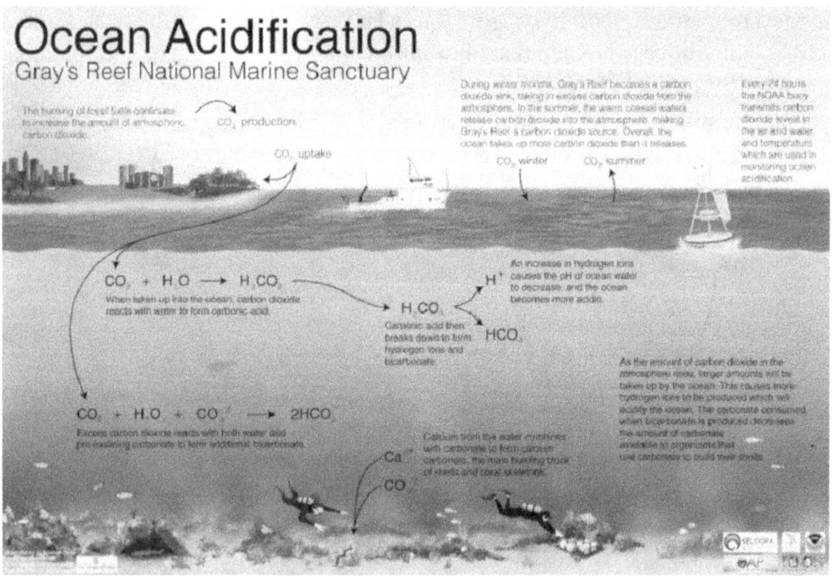

Figure 12.9 Illustration of the Chemistry Involved in Ocean Acidification. *Source*: NOAA National Marine Sanctuaries. Climate Change and Ocean Acidification, Gray's Reef. by Amanda Camp under the direction of Dr. Scott Noakes and Eugene Wright, The University of Georgia. https://sanctuaries.noaa.gov/science/sentinel-site-program/grays-reef/climate-change-ocean-acidification.html

Figure 12.10 Coral Bleaching on the Great Barrier Reef in Australia. *Source*: NOAA. Climate.gov. Unprecedented 3 years of global coral bleaching, 2014–2017. https://www.climate.gov/news-features/understanding-climate/unprecedented-3-years-global-coral-bleaching-2014%E2%80%932017

Researchers have already discovered severe levels of pteropod shell dissolution in the Southern Ocean around Antarctica.

Effects on Animals and Ecosystems

The extreme warming in the far North is particularly troubling. As ice sheets melt, many animals that depend on the presence of an ice sheet for food and shelter are losing their habitat. An example is the plight of polar bears near the Arctic Circle that are adapted to hunt seals and other prey on ice sheets. With warming temperatures, the ice sheets are melting, and the bears are unable to adapt to the loss of their food source (figure 12.11).

Another species that has been particularly hard hit by northern temperature increases are caribou (figure 12.12), which depend on the presence of lichens as a winter food source.[13] Global warming has had two effects on the caribous' winter food source. First, warmer temperatures in the North have increased the frequency of fires that burn the lichens and remove them as a winter food source.[14] Second, there is evidence that warmer weather allows the encroachment of inedible mosses and woody plants that crowd out the lichens and deprive the caribou of their food source.[15]

Ecosystem changes due to climate change are not restricted to northern climates. As global temperatures rise, many plant communities are relocating to areas that were previously inhospitable. We can measure the ideal

Figure 12.11 Starving Polar Bear Is Unable to Hunt because Ice Sheet Has Melted. *Source*: Wikimedia Commons. Andreas Weith. https://creativecommons.org/licenses/by-sa/4.0/deed.en

Figure 12.12 Caribou. They may lose their winter food source if lichens are lost. *Source*: U.S. Department of the Interior. Tackling the Climate Crisis. https://www.doi.gov/priorities/tackling-climate-crisis

growing conditions for plants by examining *plant hardiness zones*. These are geographical areas in which the conditions allow plant species (and communities) to thrive. Figure 12.13 shows the shift in plant hardiness zones as they appeared in the United States between the time period 1981–2010, compared to how they will appear between 2011 and 2040. We can see that the plant hardiness zones will shift to the north. For example, plant species that thrived in peninsular Florida from 1981 to 2010 are now found farther north into Georgia and South Carolina.

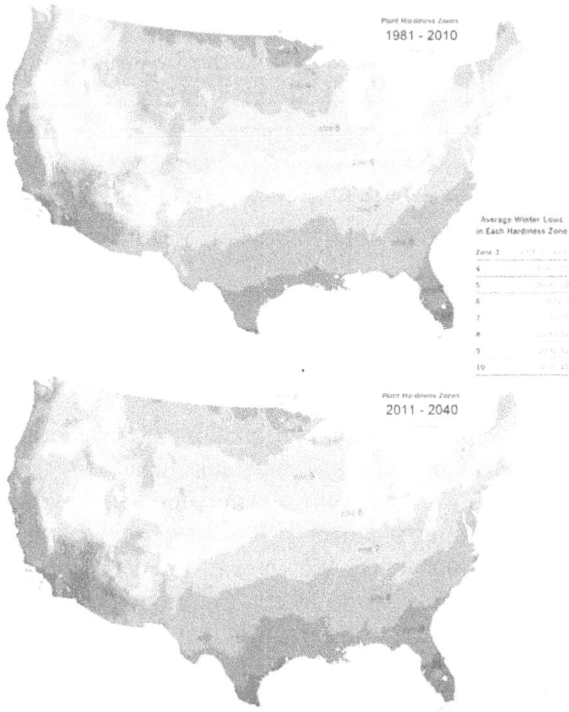

Figure 12.13 Changes in Plant Hardiness Zones, 1981–2010 and 2011–2040. *Source*: NOAA. Climate Prediction Center: Plant hardiness zones, https://www.cpc.ncep.noaa.gov/products/wesley/cfsr/plant_hardiness_zones.html

Climate change is forcing some animal species into extinction. Changes in precipitation patterns likely associated with climate change are responsible for the extinction of two populations of butterflies in southern California. The plants that were necessary for the development of butterfly larvae did not survive the weather changes, and the populations soon crashed.[16] In another example, cold-water fish species like trout and salmon, already stressed by habitat loss and pollution, cannot tolerate the increased water temperature and other changes associated with climate change.[17]

Furthermore, climate change has led to habitat destruction due to an increase in wildfires, especially in western states. A 2024 study found that intense wildfires are now twice as common worldwide as they were twenty years ago.[18] Increased fire frequency is exacerbated by antiquated forestry practices discussed in chapter 5.

Climate change is also responsible for changes in important biological timing in many species. Some seasonal biological activities are happening 15 to 20 days earlier than several decades ago, such as tree species that bloom earlier, migratory birds that arrive earlier, butterflies that emerge earlier, and terrestrial reptiles and amphibians that reproduce earlier. This can have devastating effects on important activities like reproduction. An example is the European Pied Flycatcher, a migratory bird species whose young eat caterpillars of insects that normally emerge just as the young flycatchers hatch. However, with climate change, spring arrives earlier, and the insects emerge earlier. The flycatchers now frequently arrive after the caterpillars have matured into insects, which their chicks cannot eat. Like many migratory bird species worldwide, pied flycatcher populations are declining due to climate-related food shortages.[19]

Effects on Agriculture and Fisheries

Agriculture and fisheries depend on specific climate conditions, so climate change can have a major impact. Temperature changes can cause habitat ranges and crop planting dates to shift, and droughts and floods due to climate change may hinder farming practices.

Let's examine crop plants first. We know that under the proper conditions of temperature and humidity, plants take in CO_2 from the atmosphere and use the radiant energy of the sun to convert CO_2 and water into sugars. Oxygen (O_2) is released as a waste gas. It's tempting to think, then, that increases in temperature and atmospheric CO_2 associated with climate change will be a great benefit to crop plants. But it turns out that the truth is far more complicated. Several studies have explored crop plants under greenhouse conditions where temperature and CO_2 are carefully controlled. One such study examined maize plant growth under normal temperatures and warmer (+4ºC) temperatures. The result was a slight increase in the mass of the crop plants themselves, but a dramatic decrease in crop yield.[20] The same results were found in several other crop species.[21] These studies confirm that higher temperatures associated with climate change will decrease crop yield. To make the situation even worse, increased temperatures lead to water deficits which further reduce crop yield.

But what about CO_2 levels? Studies have shown that higher concentrations of atmospheric CO_2 affect crops in two ways. First, they boost crop yields in some species by increasing the rate of photosynthesis, which spurs growth. Second, they reduce the amount of water crops lose through transpiration, which is the loss of water through pores in the leaves. These studies have found that some crops benefit while other crops suffer from increased atmospheric CO_2 levels.[22] However, any gains in crop yield from increased CO_2 levels are more than offset by damage from increased temperatures and water loss. The conclusion is that crop yield is likely to suffer from the effects of climate change.

There is also evidence that climate change can have other effects on agriculture, such as changes in the growing season. Scientific research shows that climate change has altered growing seasons across the globe. A recent example examined the grape harvest in Napa Valley, California, where researchers demonstrated that the start of the wine grape growing season now comes nearly a month earlier than it did in the 1950s because of the region's warming climate.[23] By 2100, it appears likely that Napa Valley will no longer support a grape harvest due to increased temperatures and decreased rainfall.

Climate change can make conditions either better or worse for growing crops in different parts of the earth. For example, changes in temperature, rainfall, and frost-free days are leading to longer growing seasons in almost every state.[24] A longer growing season can have both positive and negative impacts on agriculture, though. In some cases, farmers can opt to plant longer-maturing crops or more crop cycles, while in others may have to provide additional irrigation over a longer, hotter growing season.[25] In some cases, valuable croplands of today are likely to suffer from heat and drought and become less suitable for agriculture in the future.

In summary, we have seen that the effects of global climate change are very real, and that they affect many aspects of life on earth, including temperature changes, changes in the global hydrological cycle and weather, melting glaciers and sea level rise, ocean acidification, and effects on organisms and ecosystems. We will next explore the regulation of climate change through the legal system and investigate some actions being taken to minimize the effects of global climate change.

12.3 ENVIRONMENTAL LAWS AND GLOBAL CLIMATE CHANGE

Unfortunately, there is no single piece of legislation in the United States that deals with climate change in a comprehensive way. The closest single environmental law that includes climate change provisions is the Clean Air Act (CAA) that we discussed in chapter 6. Two of the six criteria pollutants subject to National Ambient Air Quality Standards, NO_x and ozone (O_3), can be considered to be greenhouse gases, as can many of the hydrocarbons that are listed as hazardous air pollutants under CAA Section 112. Recall that CO_2, a powerful and abundant greenhouse gas, must be regulated by the EPA based on a 2007 Supreme Court decision, but those regulations are still pending at this writing.

The NEPA that we discussed in chapter 1 requires that every federal agency consider the environmental consequences of its actions. Dealing with climate change is a national priority in the United States, so the effects of a proposed agency action must consider the impacts on greenhouse gas emissions, effects on temperature, and other issues related to climate change.

Several recent United States laws address various aspects of climate change.

- **Energy Policy Act of 2005:**[26] Promotes nuclear power by providing tax breaks for the nuclear power industry. In addition, the Act provides funding to extend the renewable electricity production credit and in new tax incentives for investment in clean coal facilities. Loan guarantees were granted for innovative technologies such as advanced nuclear reactors and clean coal. The Act also provided subsidies to wind energy, promoted the competitiveness of geothermal energy in relation to fossil fuels, and created a biomass grant program.
- **Energy Independence and Security Act of 2007:**[27] Supported the expansion of the production of renewable fuels, reduced U.S. dependence on oil, and increased energy security. It sets a mandatory Renewable Fuel Standard (RFS) requiring fuel producers to increase the use of biofuels and provides incentives for the development of renewable energy technologies. The act also includes provisions phasing out the use of incandescent light bulbs and improving lighting efficiency by more than 70 percent.
- **Food, Conservation, and Energy Act of 2008:**[28] Expands the Biorefinery Assistance Program by providing loan guarantee funding for the creation of commercial-scale biorefineries as well as grants to build demonstration-scale biorefineries. A 2014 revision reauthorizes and provides additional funding for energy programs. The Biorefinery Assistance Program is expanded to include bio-based products and renewable chemicals.
- **American Recovery and Reinvestment Act of 2009:**[29] Authorizes a stimulus package that supports new and existing renewable energy and energy efficiency programs.
- **Inflation Reduction Act of 2022:**[30] Along with reducing inflation, it creates a system for investing in domestic energy production while promoting clean energy.

During this time, there were several Presidential Executive Orders that addressed aspects of climate change. These Executive Orders often revoked previous Executive Orders and generally reflected the regulatory approaches of successive administrations.[31]

In addition to these federal actions, many states have developed climate change policies and laws. According to the Association of State and Territorial Health Officials (ASTHO), at least seventeen states have introduced legislation related to climate change as of 2023.[32] Of these, nine states (CA, HI, MD, MN, NM, NY, OR, RI, VT) are considering bills that direct funding for climate change-related projects and programs. Seven states (CA, CT, HI, MN, NY, OR, RI) introduced legislation to implement climate change education programs. Ten states (CA, HI, IN, MD, MN, NJ, NY, RI, TX, WA) have introduced legislation that directs state agencies to update and/or publish climate plans, assessments, and/or reports.

In the international arena, there have been several international agreements that address global climate change in various ways.

The UN Framework Convention on Climate Change (UNFCCC) established the basic legal framework and principles for international climate change cooperation with the goal of stabilizing atmospheric concentrations of greenhouse gases. To increase the effectiveness of the 1992 UNFCCC, the Kyoto Protocol (1997) committed industrialized countries and countries in transition to a market economy to achieve emissions reduction targets for six greenhouse gases. The Kyoto Protocol's first commitment period was from 2008 to 2012, but it was extended by the 2012 Doha Amendment for a second commitment period from 2013 to 2020.

In 2015, 196 parties adopted the Paris Agreement, which requires all parties to determine, plan, and regularly report on the nationally determined contribution (NDC) that they undertake to mitigate climate change. Parties also submit aggregate progress on mitigation, adaptation, and means of implementation, which are reviewed every five years. The United States has been an intermittent member of the Paris Agreement, depending on the particular administration.

12.4 SOLUTIONS TO CLIMATE CHANGE

The news constantly bombards us with reports of worsening climate catastrophes around the U.S. and the world. It's easy to conclude that the effects of global climate change are inevitable and that we are powerless to do anything about it. But there are still reasons to be hopeful.

To be sure, climate change is one of the most complex issues facing humanity. The effects of climate change involve many dimensions of civilization, including science, economics, education, politics, and moral and ethical considerations. Climate change is a global problem that is experienced on local and personal levels by everybody. We will experience the effects of global climate change for centuries, if not millennia, in the future.

The major culprit is atmospheric CO_2, the heat-trapping greenhouse gas that is the primary driver of recent climate change and global warming. Carbon dioxide remains in the atmosphere for thousands of years, and the planet (especially the ocean) takes centuries to respond. Even if we stopped emitting all greenhouse gases today, global warming and climate change will continue to affect future generations. In a sense, we are already committed to some level of climate change, even if our responses are immediate and thorough.

The degree to which we experience the effects of climate change remains to be determined. It will depend on how much global greenhouse gas emissions continue and exactly how our climate responds to those emissions. Despite increasing awareness of climate change, our emissions of greenhouse gases continue to increase. In 2013, the daily level of carbon dioxide in the atmosphere surpassed 400 parts per million for the first time in human history. The last time levels were that high was over three million years ago during the Pliocene Epoch.[33]

It's clear that we are already committed to some level of climate change, since we are already experiencing the effects. The longer it takes us to respond, the worse these effects will be. Our responses to climate change involve a two-pronged approach: mitigation and adaptation.

- **Mitigation** is reducing emissions of and stabilizing the levels of greenhouse gases in the atmosphere. It involves reducing the flow of greenhouse gases into the atmosphere, either by reducing sources of these gases (for example, the burning of fossil fuels) or enhancing the "sinks" that accumulate and store these gases (such as the oceans, forests, and soils).[34] The goal of mitigation is to avoid significant human interference with earth's climate, "stabilize greenhouse gas levels in a timeframe sufficient to allow ecosystems to adapt naturally to climate change, ensure that food production is not threatened, and to enable economic development to proceed in a sustainable manner."[35] Mitigation is the object of *emission trading programs*, particularly in CO_2.[36]
- **Adaptation** is adjusting to climate change that is already happening. Adaptation involves adjusting to actual or expected future climates. The goal is to reduce our risks from the harmful effects of climate change (like sea-level rise, more intense extreme weather events, or food insecurity). It also includes making the most of any potential beneficial opportunities associated with climate change (for example, longer growing seasons or increased yields in some regions).

Dealing with climate change is not new to humans. Our shared history has many examples of the ways that people and societies have adjusted to and coped with changes in climate and weather extremes.[37] Climate change has been at least partly responsible for the rise and fall of several civilizations. In fact, earth's climate has been relatively stable for the past 10,000 years. It is this stability that has allowed for the development of our modern civilization and agriculture. Modern life as we know it is adapted to a stable climate and not the much warmer climate with severe weather that we are facing. As our climate changes, we will have to adapt. The faster the climate changes, the more difficult it will be.[38]

Climate change is, of course, a global issue. However, it is experienced at a local level, so local governments are at the leading edge of adaptation. Cities and local communities around the world have been focusing on solving their own climate problems.[39] They are working to design and build flood defenses, improve water storage and use, improve energy infrastructures, plan for heat waves and higher temperatures, and plan for more frequent and powerful weather events.

State and local governments are also getting better at adaptation. Climate change is being included in development plans, including the management of increasingly extreme disasters, methods to protect coastlines and deal with sea-level rise, ways to manage land and forests, plans for dealing with severe weather and drought, development of new crop varieties, and ways to protect energy and public infrastructure.[40] Many state and local governments, as well as businesses, have appointed resilience officers whose job is to prepare for the effects of climate change.[41]

A report by the National Academy of Science of the United States contains a statement that summarizes the view of the United States and many other countries toward climate change. It leaves us with a message of hope, and that is a good place to conclude our discussion of global climate change:

> *The climate challenge is large and complex. But it is very likely*
> *that many people, working from many angles,*
> *can help address climate change and its*
> *ecological consequences.*[42]

* * *

Review Questions

1. Describe what is meant by climate change.
2. Explain how CO_2 levels are linked to global temperature changes.
3. Explain how global warming is more rapid in Arctic areas.
4. What is the relationship between the retreat of glaciers, climate change, and sea level rise?
5. Describe the effect of ocean acidification on bivalve animals and corals.
6. Name and describe four United States laws that address climate change.
7. What international agreements directly address global climate change?

Case Analysis

As we have seen, increases in atmospheric CO_2 levels account for many of the consequences of global climate change. Because CO_2 is not regulated as a criteria pollutant or a hazardous air pollutant, EPA had to find another way to regulate it under the Clean Air Act. It did so in 2019 with its Clean Power Plan rule, which established a "best system of emission reduction" by restructuring the United States' overall mix of electricity generation, to transition from 38 percent coal to 27 percent coal by 2030. However, the Clean Power Plan was immediately challenged by several states and power providers. The case eventually reached the United States Supreme Court. The following is an extract of Chief Justice Robert's 2022 decision for the Supreme Court.[43]

> The Clean Air Act establishes three main regulatory programs to control air pollution from stationary sources such as power plants. Clean Air Amendments of 1970, 42 U.S.C. §7401 et seq. One program is the New Source Performance Standards program of Section 111, at issue here. The other two are the National Ambient Air Quality Standards (NAAQS) program, set out in Sections 108 through 110 of the Act, 42 U.S.C. §§7408–7410, and the Hazardous Air Pollutants (HAP) program, set out in Section 112, §7412. To understand the place and function of Section 111 in the statutory scheme, some background on the other two programs is in order.
>
> . . .
>
> [I]n October 2015, EPA promulgated two rules addressing carbon dioxide pollution from power plants—one for new plants under Section 111(b), the other for existing plants under Section 111(d). Both were premised on the Agency's earlier finding that carbon dioxide is an "air pollutant" that "may reasonably be anticipated to endanger public health or welfare" by causing climate change. Carbon dioxide is not subject to a NAAQS and has not been listed as a toxic pollutant.
>
> The first rule announced by EPA established federal carbon emissions limits for new power plants of two varieties: fossil-fuel-fired electric steam generating units (mostly coal fired) and natural-gas-fired stationary combustion turbines. Following the statutory process set out above, the Agency determined the BSER [best system of emission reduction . . . adequately demonstrated] for the two categories of sources.
>
> The second rule was triggered by the first: Because EPA was now regulating carbon dioxide from new coal and gas plants, Section 111(d) required EPA to also address carbon emissions from existing coal and gas plants. See §7411(d)(1). It did so through what it called the Clean Power Plan rule.
>
> . . .
>
> [T]he Agency replaced the Clean Power Plan by promulgating a different Section 111(d) regulation, known as the Affordable Clean Energy (ACE) Rule. Based on its view of what measures may permissibly make up the BSER, EPA determined that the best system would be akin to building block one of the

Clean Power Plan: a combination of equipment upgrades and operating practices that would improve facilities' heat rates. The ACE Rule determined that the application of its BSER measures would result in only small reductions in carbon dioxide emissions.

. . .

In devising emissions limits for power plants, EPA first "determines" the "best system of emission reduction" that—taking into account cost, health, and other factors—it finds "has been adequately demonstrated." 42 U. S. C. §7411(a)(1). The Agency then quantifies "the degree of emission limitation achievable" if that best system were applied to the covered source. The BSER, therefore, "is the central determination that the EPA must make in formulating [its emission] guidelines" under Section 111. The issue here is whether restructuring the Nation's overall mix of electricity generation, to transition from 38% coal to 27% coal by 2030, can be the "best system of emission reduction" within the meaning of Section 111.

. . .

Under our precedents, this is a major questions case. In arguing that Section 111(d) empowers it to substantially restructure the American energy market, EPA "claim[ed] to discover in a long-extant statute an unheralded power" representing a "transformative expansion in [its] regulatory authority." It located that newfound power in the vague language of an "ancillary provision[ ]" of the Act, one that was designed to function as a gap filler and had rarely been used in the preceding decades. And the Agency's discovery allowed it to adopt a regulatory program that Congress had conspicuously and repeatedly declined to enact itself. Given these circumstances, there is every reason to "hesitate before concluding that Congress" meant to confer on EPA the authority it claims under Section 111(d).

. . .

But, the Agency explained, in order to "control[ ] CO_2 from affected [plants] at levels . . . necessary to mitigate the dangers presented by climate change," it could not base the emissions limit on "measures that improve efficiency at the power plants." "The quantity of emissions reductions resulting from the application of these measures" would have been "too small." Instead, to attain the necessary "critical CO_2 reductions," EPA adopted what it called a "broader, forward-thinking approach to the design" of Section 111 regulations. Rather than focus on improving the performance of individual sources, it would "improve the overall power system by lowering the carbon intensity of power generation." Ibid. And it would do that by forcing a shift throughout the power grid from one type of energy source to another. In the words of the then-EPA Administrator, the rule was "not about pollution control" so much as it was "an investment opportunity" for States, especially "investments in renewables and clean energy."

This view of EPA's authority was not only unprecedented; it also effected a "fundamental revision of the statute, changing it from [one sort of] scheme of . . . regulation" into an entirely different kind. Under the Agency's prior view of Section 111, its role was limited to ensuring the efficient pollution performance of each individual regulated source. Under that paradigm, if a source was already operating at that level, there was nothing more for EPA to do. Under its newly "discover[ed]" authority, however, EPA can demand much greater reductions in emissions based on a very different kind of policy judgment: that it would be "best" if coal made up a much smaller share of national electricity generation. And on this view of EPA's authority, it could go further, perhaps forcing coal plants to "shift" away virtually all of their generation—i.e., to cease making power altogether.

‘ ‘ ‘

There is little reason to think Congress assigned such decisions to the Agency. For one thing, as EPA itself admitted when requesting special funding, "Understand[ing] and project[ing] system-wide . . . trends in areas such as electricity transmission, distribution, and storage" requires "technical and policy expertise not traditionally needed in EPA regulatory development." "When [an] agency has no comparative expertise" in making certain policy judgments, we have said, "Congress presumably would not" task it with doing so.

We also find it "highly unlikely that Congress would leave" to "agency discretion" the decision of how much coal- based generation there should be over the coming decades. The basic and consequential tradeoffs involved in such a choice are ones that Congress would likely have intended for itself. Congress certainly has not conferred a like authority upon EPA anywhere else in the Clean Air Act. The last place one would expect to find it is in the previously little-used backwater of Section 111(d).

. . .

Capping carbon dioxide emissions at a level that will force a nationwide transition away from the use of coal to generate electricity may be a sensible "solution to the crisis of the day." But it is not plausible that Congress gave EPA the authority to adopt on its own such a regulatory scheme in Section 111(d). A decision of such magnitude and consequence rests with Congress itself, or an agency acting pursuant to a clear delegation from that representative body. The judgment of the Court of Appeals for the District of Columbia Circuit is reversed, and the cases are remanded for further proceedings consistent with this opinion.

Questions

1. What was the result of the Supreme Court's decision? Will the EPA's Clean Power Plan go into effect?
2. The Court rejected the EPA's argument that restructuring the United States' overall mix of electricity generation from coal to other sources was appropriate. What was the basis of the Court's decision?
3. What is the "major questions" doctrine the Court describes? Could it have an effect on other attempts to regulate greenhouse gases?

4. Did the Supreme Court say that the EPA could not regulate CO_2 at all? What other options might be available to the EPA?
5. What does this Supreme Court decision say about possible future attempts by federal agencies to regulate greenhouse gases?

NOTES

1. Temperatures may be rising even faster. Global temperatures in 2023 rose about 2.1°F (1.2°C) above the average for NASA's baseline period (1951–1980). https://www.nasa.gov/news-release/nasa-analysis-confirms-2023-as-warmest-year-on-record/.

2. NOAA. *Climate Change Impacts.* https://www.noaa.gov/education/resource-collections/climate/climate-change-impacts.

3. See NASA. *Extreme Weather and Climate Change.* https://science.nasa.gov/climate-change/extreme-weather/.

4. NOAA. *2023: A Historic Year of U.S. Billion-Dollar Weather and Climate Disasters.* https://www.climate.gov/news-features/blogs/beyond-data/2023-historic-year-us-billion-dollar-weather-and-climate-disasters.

5. United Nations Convention to Combat Desertification (UNCCD). 2024. *The Global Threat of Drying Lands: Regional and Global Aridity Trends and Future Projections.* UNCCD.

6. NASA. *Slowdown of the Motion of the Ocean.* https://science.nasa.gov/earth/earth-atmosphere/slowdown-of-the-motion-of-the-ocean/.

7. NOAA. Climate.gov. *Climate Change: Global Sea Level.* https://www.climate.gov/news-features/understanding-climate/climate-change-global-sea-level.

8. See the excellent discussion at NOAA Education. *Ocean Acidification.* https://www.noaa.gov/education/resource-collections/ocean-coasts/ocean-acidification.

9. Recall that pH is a measure of the relative acidity or alkalinity of a solution. The pH scale is logarithmic and runs from 0 to 14. Neutral pH is 7, with numbers above 7 being increasingly alkaline (basic) and numbers below 7 being increasingly acidic.

10. Fischetti, M. 2012. Ocean Acidification Can Mess with a Fish's Mind. *Scientific American*, September 27. https://www.scientificamerican.com/article/ocean-acidification-can-m/.

11. See, for example, Anthony, K. R., Kline, D. I., Diaz-Pulido, G., Dove, S., & Hoegh-Guldberg, O. 2008. Ocean Acidification Causes Bleaching and Productivity Loss in Coral Reef Builders. *Proceedings of the National Academy of Sciences, 105*(45), 17442–17446.

12. Bednaršek, N., Tarling, G., Bakker, D., et al. 2012. Extensive Dissolution of Live Pteropods in the Southern Ocean. *Nature Geoscience, 5*, 881–885. https://doi.org/10.1038/ngeo1635.

13. Caribou, also known as reindeer (*Rangifer tarandus*), are a large species of deer with a circumpolar distribution. In many parts of their range, caribou depend on

lichens as a food source. Lichens are a symbiotic partnership of two separate organisms, a fungus and an alga.

14. U.S. Geological Survey. 2014. *Climate Change Could Alter Range of Caribou and May Impact Hunters' Access*. https://www.usgs.gov/news/national-news-release/climate-change-could-alter-range-caribou-and-may-impact-hunters-access.

15. See, for example: He, L., Chen, W., Fraser, R. H., Schmelzer, I., Arsenault, A., Leblanc, S. G., . . . & Brodeur, A. 2024. Satellite-Detected Decreases in Caribou Lichen Cover, *Cladonia* (*Cladina*) spp., over Eastern Canada During the Last Three Decades. *Forest Ecology and Management, 556*, 121753.

16. McLaughlin, J.F., Hellmann, J.J., Boggs, C.L., & Ehrlich, P.R. 2002. Climate change hastens population extinctions. *Proceedings of the National Academy of Sciences of the United States of America,* 99(9), 6070–6074. https://doi.org/ 10.1073/pnas.052131199.

17. Williams, J.E., Isaak, D.J., Imhof, J., Hendrickson, D.A., & McMillan, J.R. 2015. *Cold-Water Fishes and Climate Change in North America, Reference Module in Earth Systems and Environmental Sciences*. Elsevier. https://doi.org/10.1016/B978-0-12-409548-9.09505-1.

18. Cunningham, C.X., Williamson, G.J., & Bowman, D.M.J.S. 2024. Increasing Frequency and Intensity of the Most Extreme Wildfires on Earth. N*ature Ecology and Evolution*. https://doi.org/10.1038/s41559-024-02452-2.

19. Lamers, K.P., Nilsson, J.A., Nicolaus, M., & Both, C. 2023. Adaptation to Climate Change Through Dispersal and Inherited Timing in an Avian Migrant. *Nature Ecology & Evolution, 7*(11), 1869–1877. See also Bange, K., Friggens, M., & Finch, D. 2011. *A System for Assessing Vulnerability of Species (SAVS) to Climate Change*. General Technical Report RMRS-GTR-257. USDA Rocky Mountain Research Station, p. 37.

20. Hatfield, J.L., & Prueger, J.H. 2015. Temperature Extremes: Effect on Plant Growth and Development. *Weather and Climate Extremes, 10*, Par(A), 4–10. https://doi.org/10.1016/j.wace.2015.08.001.

21. Zhao, C., Liu, B., Piao, S., Wang, X., Lobell, D. B., Huang, Y., . . . & Asseng, S. 2017. Temperature Increase Reduces Global Yields of Major Crops in Four Independent Estimates. *Proceedings of the National Academy of sciences, 114*(35), 9326–9331; and Mohammadi, S., Rydgren, K., Bakkestuen, V., & Gillespie, M.A.K. 2023. Impacts of Recent Climate Change on Crop Yield Can Depend on Local Conditions in Climatically Diverse Regions of Norway. *Scientific Reports, 13,* 3633. https://doi.org/10.1038/s41598-023-30813-7.

22. See Deryng, D., Conway, D., Ramankutty, N., Price, J., & Warren, R. 2014. Global Crop Yield Response to Extreme Heat Stress Under Multiple Climate Change Futures. *Environmental Research Letters, 9*(3), 034011.

23. Cayan, D., DeHaan, L., Tyree, M., and Nicholas, K.A. 2023. A 4-Week Advance in the Growing Season in Napa Valley, California, USA. *International Journal of Climatology, 43*(12), 5586–5603.

24. See Gowda, P., Steiner, J.L., Olson, C., Boggess, M., Farrigan, T., and Grusak, M.A. 2018. Ch. 10: Agriculture and Rural Communities. In *Impacts, Risks, and Adaptation in the United States: Fourth National Climate Assessment, Volume II.*

U.S. Global Change Research Program, pp. 391–437. https://nca2018.globalchange.gov/chapter/10/.

25. See Nolte, C.G., Spero, T.L., Bowden, J.H., Mallard, M.S., & Dolwick, P.D. 2018. The Potential Effects of Climate Change on Air Quality across the Conterminous U.S. at 2030 Under Three Representative Concentration Pathways. *Atmospheric Chemistry and Physics, 18*(20), 15471–15489.

26. 42 U.S.C. §15801 et seq.

27. 42 U.S.C. §17001 et seq.

28. 42 U.S.C. §7101 et seq.

29. Pub. L. No. 111–5, 123 Stat. 115.

30. Pub. L. No 117–169, 136 Stat. 1818.

31. For example, Executive Order 13990 (2023) *Climate Crisis; Efforts to Protect Public Health and Environmental and Restore Science* (Biden Administration) revoked important sections of Executive Order 13834 (2018) *Efficient Federal Operations* (Trump Administration).

32. Association of State and Territorial Health Officials (ASTHO). 2023. *States Can Lead on Mitigating the Effects of Climate Change.* https://www.astho.org/communications/blog/states-can-lead-on-mitigating-the-effects-of-climate-change/.

33. NASA Explore. *Responding to Climate Change.* https://science.nasa.gov/climate-change/adaptation-mitigation/.

34. See EPA. *Inventory of U.S. Greenhouse Gas Emissions and Sinks.* EPA 430-R-20–002. https://www.epa.gov/ghgemissions/inventory-us-greenhouse-gas-emissions-and-sinks-1990-2018.

35. International Panel on Climate Change (IPCC). *AR5 Climate Change 2014: Mitigation of Climate Change.* https://www.ipcc.ch/report/ar5/wg3/.

36. Emissions trading programs consist of a limit (or cap) on pollution and tradable allowances equal to the limit that authorize allowance holders to emit a specific quantity (e.g., one ton) of the pollutant. This limit ensures that the environmental goal is met, and the tradable allowances provide flexibility for individual emissions sources to set their own compliance path. Because allowances are bought and sold in an allowance market, these programs are often referred to as "market based." See EPA. *Emissions Trading Resources. What Is Emissions Trading?* https://www.epa.gov/emissions-trading-resources/what-emissions-trading.

37. See, for example: Steel, D., DesRoches, C. T., & Mintz-Woo, K. 2022. Climate Change and the Threat to Civilization. *Proceedings of the National Academy of Sciences, 119*(42), e2210525119; and Fagan, B. 2008. *The Great Warming: Climate Change and the Rise and Fall of Civilizations.* Bloomsbury Publishing.

38. NASA Explore. *Responding to Climate Change.* https://science.nasa.gov/climate-change/adaptation-mitigation/.

39. See Salvador Costa, M.J., Leitão, A., Silva, R., Monteiro, V., & Melo, P. 2022. Climate Change Prevention through Community Actions and Empowerment: A Scoping Review. *International Journal of Environmental Research and Public Health,* 19(22), 14645. https://doi.org/ 10.3390/ijerph192214645.

40. See IPCC. 2014. *Climate Change 2014: Impacts, Adaptation, and Vulnerability. Part A: Global and Sectoral Aspects. Contribution of Working Group II to*

the Fifth Assessment Report of the Intergovernmental Panel on Climate Change. Cambridge University Press, pp. 1–32.

41. The National Conference of State Legislators (NCSL) maintains a website with information for state resilience officers at https://www.ncsl.org/environment-and-natural-resources/state-resilience-offices. Many local and state governments, as well as business and industry organizations, maintain resources that support resiliency practices and officers.

42. National Research Council, Division on Earth, Life Studies, Board on Life Sciences, & Committee on Ecological Impacts of Climate Change. 2008. *Ecological Impacts of Climate Change*. National Academies Press. https://nap.nationalacademies.org/catalog/12491/ecological-impacts-of-climate-change.

43. *West Virginia v. Environmental Protection Agency*, 597 U.S. 697, 142 S. Ct. 2587 (2022).

Glossary of Terms

Abyssopelagic Zone
Extends from 4,000-6,000m deep in the open ocean, commonly known as "The Abyss" due to the general absence of life found at these great depths. There is no light and little oxygen at these depths.

Acid Rain
Occurs when sulfur dioxide (SO_2) or nitrogen oxides (NO_x) react with water in the atmosphere to form sulfuric or nitric acids, which then fall to the surface of the earth as acidic precipitation.

Adaptation
As used in climate change, the process of adjusting to climate change that is already happening. Involves adjusting to actual or expected future climates. The goal is to reduce risks from the harmful effects of climate change and make the most of beneficial opportunities associated with climate change.

Agent Orange
A commercial and military defoliant, it is a mixture of the herbicides 2,4,5-T and 2,4-D, which was used by the U.S. military against North Vietnamese forces who used the forests for cover.

Aquifer
See Groundwater.

Assimilation
The process by which organic nitrogen compounds form from inorganic nitrogen compounds in an ecosystem.

Atmosphere
The layer of gases that surrounds the earth that we call "air." It extends from chambers below the earth's surface to more than 10,000km above the surface.

Autotroph
A plant that uses energy from the sun to produce glucose for food through the process of photosynthesis.

Bathypelagic Zone
Also known as the "midnight zone" of the open ocean, it lies between 1,000 and 4,000m deep. Light cannot penetrate to these depths, leaving this region of the ocean in complete darkness.

Benthic Zone
The bottom sediments of a lake where a diverse variety of activities take place may be made up of sand, mud, marl, and/or a mixture of any or all of these materials.

Biodiversity
The number of different species of organisms (plants and animals) in an ecosystem.

Bioenergy
See Biomass Energy

Biogas
Gas, usually methane, that is produced in anaerobic digesters at sewage treatment plants and at dairy and livestock operations. It also forms in and may be captured from solid waste landfills. Also known as biomethane or renewable natural gas,

Biogeochemical Cycle
Elements (like nitrogen or carbon) and compounds (such as water or carbon dioxide) move between the biotic and abiotic components of an ecosystem.

Biomass
As used in ecology means the mass of living organisms in a given area or ecosystem at a given time, and is expressed in units of weight (grams, kg, etc.).

Biomass energy
Energy derived from renewable organic material from plants and animals. Also known as bioenergy.

Biosphere
All living things on the planet, which includes all of the microorganisms, plants, and animals of earth.

Brownfields
Areas where past contamination of land means that new establishments will not relocate to a contaminated site because of potential liability under CERCLA.

Carnivore
An organism that consumes other organisms.

Carrying Capacity
The maximum number of individuals (N) of a species that can be supported by a particular environment.

Clear-Cutting
Forestry technique that removes most, if not all, trees from an area and is often coupled with replanting. The resulting trees are then all the same age.

Climate
Atmospheric changes over longer periods of time, usually thirty years or more.

Conservationist Movement
The view that careful management of the environment would ensure plentiful natural resources for the future.

Conventional Pollutants (Clean Water Act)
Pollutants traditionally regulated as discharges from POTWs, such as biological oxygen demand (BOD), suspended solids, fecal coliform bacteria, pH, oil, and grease.

Defendant
The party against whom a lawsuit is brought.

Denitrification
The process by which nitrates are returned to the atmosphere as nitrogen gas by denitrifying bacteria in soils.

Ecological Niche
The position of a species within an ecosystem, comprising both the habitat requirements and the functional role of a species.

Ecologically Sustainable Forestry
An approach to removing trees from forests in ways that do not unduly affect the viability of other trees. Seen as the most environmentally sound forestry technique.

Effluent
Liquids discharged by factories and other structures in their discharged waters.

Endangered Species
A species of animal or plant that is in danger of extinction.

Environmental Assessment (EA)
A brief statement regarding environmental effects of an agency action under NEPA.

Environmental Impact Statement (EIS)
A detailed statement prepared by a federal agency when a proposed action will have significant environmental effects under NEPA.

Epipelagic Zone
The "sunlight zone" of the open ocean, from the surface to 200m deep. Light can penetrate through this layer so photosynthesis can take place, which supports a thriving algae community.

Estuary
Transition area between a river and the ocean, are under tidal influence, so the depth of water fluctuates depending on the tide. Protected from the full force of ocean waves, winds and storms by landforms such as barrier islands or peninsulas.

Euphotic Zone
The area of a lake or pond from the surface down to the depth where light becomes too low for organisms to undergo photosynthesis. Includes a Littoral Zone and a Limnetic Zone.

Eutrophic
Lakes or ponds that are high in nutrients and contain large populations of aquatic plants, algae, and fish. Aquatic plants and algae often grow to nuisance levels, and the fish species are generally tolerant of warm temperatures and low oxygen conditions.

Evapotranspiration or **Transpiration**
Water that evaporates from the surface of leaves of plants that drives the internal circulation of the plant.

Exponential Growth
A population will grow at a slow rate early in time, but the rate will rapidly increase to a maximum. Often seen in population growth of organisms in a new environment.

Externality
A side effect or consequence of an industrial or commercial activity that affects other parties without this being reflected in the cost of the goods or services involved. May be negative or positive.

Floodplain
The low, flat land spreading out from the downstream channel of a river or stream, periodically flood during heavy rains and snow melt. May remain wet for long periods of time, creating rich wetland habitats.

Food Chain
Diagram that shows the linear flow of energy from one set of organisms to another in an ecosystem.

Food Web
A representation of the interactions between the food chains in an ecosystem that represents the flow of nutrients and energy, also known as a consumer-resource system.

Fossil Fuels
Fuels from sources that are formed by natural processes over millions of years in which organic material from dead plants and animals was gradually converted by high temperature and pressure into the material that we find today. The major fossil fuels are coal, petroleum (oil), and natural gas.

Fracking
A process to extract gas from rock, especially shale. With this method, water, sand, and a mix of chemicals are forced under pressure into horizontally drilled wells, causing the shale to crack and release natural gas or oil.

Generator of Hazardous Waste
Any person who produces a hazardous waste either listed or characteristic (identified).

Geosphere
All rocks, sediments and soils, surface landforms and the processes that shape the earth's surface, as well as other materials that make up the earth's crust (surface). It also includes the semi-solid materials beneath the crust, and the liquid materials near the center of the earth.

Greenhouse Effect
Greenhouse gases in the atmosphere trap radiant heat from the sun and prevent the escape of heat into outer space. Important in maintaining warmth at the surface of the earth, it can increase global temperatures to dangerous levels when greenhouse gases increase above normal levels. See Greenhouse Gases.

Greenhouse Gases
Gases in the atmosphere that contribute to the greenhouse effect. May be natural or anthropogenic. Common greenhouse gases include CO_2, methane, and water. See Greenhouse Effect.

Groundwater
Water that sinks through soil and permeable rock formations to underground layers of sand, gravel, and porous rock called Aquifers.

Habitat
All the physical, chemical and biological attributes that affect or sustain the organisms within an ecosystem,

Hadopelagic Zone
Deepest part of the open ocean, characterized by deep sea canyons and trenches exceeding depths of 6,000m. Includes the famous Mariana Trench with the deepest known point on the E,,arth's surface, the Challenger Deep, with a depth of 11034 meters.

Harass (Endangered Species Act)
An intentional or negligent act or omission which creates the likelihood of injury to wildlife by annoying it to such an extent as to significantly disrupt normal behavioral patterns which include, but are not limited to, breeding, feeding, or sheltering. [ESA regulations]

Hazardous Waste
A waste with properties that make it dangerous or capable of having a harmful effect on human health or the environment.

Herbivore
An organism that consumes plants.

Heterotroph
An herbivore or carnivore that consumes plants and animals.

Hydrosphere
The liquid, solid, and gaseous water of the planet. Ranging from 10 to 20 kilometers in thickness, the hydrosphere extends from the Earth's surface down several kilometers into the lithosphere and upward about 12 kilometers into the atmosphere.

Injunction / Injunctive Relief
A legal remedy which prevents a party from doing certain acts or requires them to perform in a particular way.

intertidal Zone
An area of the ocean between the low and high tide lines (between mean high water and low water marks).

Joint and Several Liability
A common-law action in which all defendants are responsible for an act, and for the damages. The plaintiff, however, may seek to collect the awarded damages from all of the defendants (*joint liability*), or from some or only one (*several liability*) if they choose.

Keystone Species
A species that exerts a large, stabilizing influence throughout an ecosystem, despite its relatively small numerical abundance.

Lentic
The term for water bodies with little or no regular flow like lakes and ponds.

Limnetic Zone
Part of the euphotic zone of a lake or pond, the area of open water where the lake is too deep to allow rooted plant growth, contains free-floating organisms such as phytoplankton and zooplankton.

Littoral Zone
Part of the euphotic zone of a lake or pond, the nearshore area that is the shallow and warmest part of the lake where sunlight penetrates all the way

to the bottom sediments and allows rooted aquatic plants (*macrophytes)* and attached algae to grow.

Logistic Growth
Seen in population growth curves. A population grows rapidly, but as growth approaches the carrying capacity over time, the rate of increase in growth begins to decrease, producing a characteristic sigmoid (S-shaped) growth curve.

Lotic
The term for water bodies regular flow like rivers and streams.

Macrofauna
Organisms in an ecosystem that are greater than 1 mm in size. Often refers to beach ecosystems.

Macroplastics
Plastic pollutants ≥ 25mm.

Marsh
A wetland ecosystem dominated by grasses (for example, cattail marsh).

Maximum (or optimum) sustainable yield (MSY)
The level of effort (LOE) that maximizes the difference between total revenue and total cost in a system. Used in fisheries and agriculture management.

Meiofauna
Organisms in an ecosystem that are less than 1 mm in size. Often refers to beach ecosystems.

Mesoplastics
Plastic pollutants between 5 and 25mm.

Mesotrophic
Lakes or ponds that have moderate amounts of nutrients, and contain healthy, diverse populations of aquatic plants, algae, and fish.

Microplastics
Plastic pollutants between 1 and 5mm.

Mini-Microplastics
Plastic pollutants between 1µm and 1mm.

Mitigation
As used in climate change, the process of reducing emissions of and stabilizing the levels of greenhouse gases in the atmosphere. Involves reducing the flow of greenhouse gases into the atmosphere, either by reducing sources of these gases or enhancing the "sinks" that accumulate and store these gases.

Mutualism
A relationship between two species in which both species benefit.

National Biological Survey (NBS)
Originated in 1993 under the auspices of the U.S. Fish and Wildlife Service (USFWS) to assess the condition of biodiversity in the United States, and to recommend measures to reverse the trend in lost biodiversity.

National Priorities List (NPL)
Under CERCLA, a list of the sites that require remedial action. EPA only conducts remedial actions at sites on the NPL. Sites on the NPL are listed in order of priority based on relative risk.

Nexus
A connection.

Nitrification
A two-part microbial process in which ammonium ions are converted into nitrates and nitrites.

Nitrogen fixation
The process by which nitrogen gas (N_2) is transformed into ammonium (NH_4^+), a form of nitrogen that can be used by plants.

Nonconventional Pollutants (Clean Water Act)
Pollutants that are neither toxic nor conventional, such as ammonia, chlorides, nitrates, and color.

Non-Renewable Energy
Energy from finite sources that cannot be regenerated within a human lifetime, and will eventually run out like oil, coal, and natural gas.

Oligotrophic
Lakes or ponds that are generally clear and cold, lake substrate is typically firm and sandy. Nutrient levels are low, so the lake generally does not support large populations of aquatic plants, animals, or algae.

Pelagic Zone
The open ocean, seaward of the Neritic Zone.

Photochemical Smog
Smog produced when sunlight reacts with nitrogen oxides and at least one volatile organic compound (VOC) in the atmosphere.

Photosynthesis
In plants water combines with carbon dioxide to produce sugars and other carbohydrates.

Photovoltaic (PV) Solar Energy
A technology to generate power by using devices that absorb energy from sunlight and convert it into electrical energy through semiconducting materials. Commonly known as solar cells and solar panels.

Phytoplankton
Microscopic algae, use photosynthesis as the primary producers for most aquatic food webs.

Plaintiff
The party bringing a lawsuit.

Point Source (Clean Water Act)
Any discernible, confined, and discrete conveyance, including but not limited to any pipe, ditch, channel, tunnel, conduit, well . . . or vessel or other floating craft, from which pollutants are or may be discharged. CWA §502(14)

Polyps
Tiny coral animals primarily responsible for building reefs.

Preservationist Movement
A form of natural morality that supported the preservation of large tracks of pristine wilderness.

Primary Consumer
See Herbivore

Primary Producer
See Autotroph

Procedural NEPA
Section 102 of the National Environmental Policy Act.

Profundal zone
The area in a deep lake where no light reaches, so photosynthesis is not possible. Bacteria use oxygen to break down organic matter.

Remediation
The process of removing past contamination of a site by pollutants. CERCLA is the primary federal statute for hazardous waste remediation,

Renewable Energy
Energy from unlimited, naturally replenished resources, like the sun, wind, and water.

Renewable Energy
Energy from unlimited, naturally replenished resources, like the sun, wind, and water.

Retroactive Liability
A responsible party (RP) can be held liable for hazardous waste release occurring prior to 1980 (before CERCLA was enacted). In essence, this means that a responsible party can be held liable for an action that was not Illegal at the time that it originally occurred.

Riparian Zone
The biological habitat in the immediate vicinity of a stream or river.

Secondary Consumer
See Carnivore

Segmentation
An unlawful practice under NEPA of dividing large agency actions into smaller ones to avoid cumulative impacts.

Selective Cutting (Selection Cutting)
Forestry technique in which single trees or small numbers of trees are harvested. The resulting forest consists of trees of varying ages. Seen as an improvement over clear-cutting.

Smog
A combination of harmful pollutants (often appearing relatively low to the ground as a yellow-brown haze) that are introduced into the atmosphere by both natural and human induced processes. May be sulfurous or photochemical smog.

Solar Cell
See Photovoltaic (PV) Solar Energy.

Solar Panel
See Photovoltaic (PV) Solar Energy.

Strict Liability
A common-law action in which a person is legally responsible for the consequences of an action even in the absence of fault or criminal intent on the part of the defendant.

Substantive NEPA
Section 101 of the National Environmental Policy Act.

Sulfurous Smog
Smog primarily as a product of burning large quantities of high-sulfur coal. Also called Sulfur or London-type smog.

Swamp
A wetland ecosystem dominated by trees (for example, a cypress swamp).

Take (Endangered Species Act §3)
To harass, harm, pursue, hunt, shoot, wound, kill, trap, capture, or collect [a protected species], or to attempt to engage in any such conduct.

Threatened Species
A species of animal or plant that is likely to become an endangered species in the foreseeable future.

Toxic Pollutants (Clean Water Act)
Substances with characteristics based on consideration of toxicity, persistence, degradability, and effects on organisms.

Transpiration
See Evapotranspiration.

Transporter of Hazardous Waste
Individuals or entities that move hazardous waste from one site to another by highway, rail, water, or air.

Treatment, Storage, and Disposal (TSD) facility of Hazardous Waste
Sites for long term disposal of hazardous wastes. TSD facilities are heavily regulated, and designed to avoid any leakage of hazardous wastes.

Trophic Pyramid
Shows the flow of energy through ecosystems, each step represents a trophic level, beginning with primary producers at the bottom. The width of each step represents the rate of energy flow through each trophic level.

Troposphere
The layer of the atmosphere where most weather and climate occur, composed of the atmospheric gases that we call "air."

Ultraplankton
Microscopic photosynthetic bacteria in the open ocean.

Weather
The conditions of the atmosphere over a short period of time, usually measured in hours or days.

Wind Turbines
Mechanisms operated by blades that move with the wind, can be used to spin a turbine to create energy.

Zooplankton
Small to microscopic animals, secondary consumers, single-celled to large invertebrates like fish larvae and jellyfish,

Zooxanthellae
Photosynthetic algae with a symbiotic relationship with coral polyps.

Index

About the Author

Dr. James F. Berry is a biologist, college professor, and environmental attorney. He is an adjunct professor of biology at The Citadel in Charleston, SC, and a professor emeritus of biology at Elmhurst University in Elmhurst, Illinois. During his fifty-year teaching career, he has taught many courses in biology and law at several institutions. As an attorney, Dr. Berry specialized in environmental, land use, and natural resources law. He is currently retired from active law practice, but he has practiced law in state courts in Florida and Illinois, as well as in several federal district courts and courts of appeals.

Dr. Berry's education in biology includes BS and MS degrees from Florida State University and a PhD from the University of Utah. He received his law degree from IIT Chicago-Kent College of Law, having completed the law school's Program in Environmental and Energy Law.

He is a member of many scientific and legal professional organizations and has published over one hundred scientific and legal articles, book chapters, and coauthored *Wetlands: Guide to Science, Law, and Technology* and *The Environmental Law and Compliance Handbook* (McGraw-Hill), both with Mark S. Dennison.

www.ingramcontent.com/pod-product-compliance
Lightning Source LLC
LaVergne TN
LVHW010341220926
848484LV00004BB/1076